Akbar John
Zaima Azira Zainal Abidin
Ahmed Jalal Khan Chowdhury

Bioprospects of Coastal Ecosystem and Sustainable Resource Management

AF191033

Akbar John
Zaima Azira Zainal Abidin
Ahmed Jalal Khan Chowdhury

Bioprospects of Coastal Ecosystem and Sustainable Resource Management

Noor Publishing

Cover image: www.ingimage.com

Publisher:
Noor Publishing
is a trademark of
Dodo Books Indian Ocean Ltd. and OmniScriptum S.R.L publishing group

120 High Road, East Finchley, London, N2 9ED, United Kingdom
Str. Armeneasca 28/1, office 1, Chisinau MD-2012, Republic of Moldova, Europe
Managing Directors: Ieva Konstantinova, Victoria Ursu
info@omniscriptum.com

Printed at: see last page
ISBN: 978-620-2-79106-9

BIOPROSPECTS OF COASTAL

ECOSYSTEM AND SUSTAINABLE

RESOURCE MANAGEMENT

Akbar John
Zaima Azira Zainal Abidin
Ahmed Jalal Khan Chowdhury

TABLE OF CONTENT

PREFACE

As we are about to step into the post COVID-19 pandemic time, many challenges are to be addressed, especially on the economic recovery action plan (post-COVID-19) through the sustainable utilization of natural resources and implementation of appropriate measurement practices. In order to achieve the 'Agenda 2030' of the United Nations' sustainable development goals (SDGs), natural resources need to be wisely explored. Knowledge on the coastal ecosystem, its dynamics and bioprospecting potentials are well addressed on the global scale. However, at the regional frame, potentials of the coastal ecosystem are less explored due to the complexity of the resource partitioning system and intertwined nature of multi-stakeholder intervention in decision making. Malaysia has a total coastline length of about 4809 km (divided in to 1,972 km in Peninsular Malaysia and 2837 km in East Malaysia) that has special socio-economic significance. Many strategic action plans were implemented to protect the coastline from fragmentation and degradation due to natural and manmade causes.

Coastal ecosystems are the most productive and valuable landscape constantly changing due to various environmental pressures and urbanization, they are always addressed together as 'estuarine and coastal ecosystems' (ECEs) due to its intricacy in providing ecological services. In order to interconnect the coastal ecosystem dynamics and explore its bioprospecting potentials, this book was intended to address the holistic importance of the coastal ecosystem and its bio-prospect potentials. The book is the comprehensive collection of research-based data from the studies on coastal ecosystems of Malaysia (especially from the east coast of peninsular Malaysia). The book consists of nine chapters addressing the issues related to (but not limited to) the bio-prospect potential such as screening of actinomycetes from the coastal ecosystem, microbial bioprospecting using 'omics' approach, importance of integrated Multi-trophic Aquaculture, biotic diversity and shoreline erosion in coastal ecosystem. We are optimistic to say that the in-depth knowledge and scientific insights shared in this book will contribute towards sustainable development goals holistically and in particular on SDG 13,14 and 15.

The nine chapters addressed in the present book titled '*Bioprospects of coastal ecosystem and sustainable resource management*' is authored by more than 30 researchers from various disciplines indicating the transdisciplinary knowledge offered in this book. Readers will get exposed to new knowledge in each chapter and the arrangements of all nine chapters flows with the core subject of the book. The chapters addressed in this book are 1) Seasonal variations of fish diversity and species richness at the coastal water, Pekan, Pahang, Malaysia, 2) Study of glucose-6-phosphate dehydrogenase activity assay in mangrove streptomyces for actinohordin and undercylprodigiosin production, 3) Cultivation vs the 'Omics' approach for microbial bioprospecting In The 21st Century: coastal environment in Malaysia, 4) Open water integrated multi-trophic aquaculture (IMTA) in coastal ecosystem: the status and prospects in Malaysia, 5) Antioxidant properties Of (*Nerita articulata*) drom estuarine mangrove Kuantan, Pahang Malaysia, 6) Heavy metal resistant bacteria from marine sediment of pantai Balok, Pahang, Malaysia, 7) Salinity tolerance and growth performance of Asian seabass (*Lates calcarifer*) juveniles, 8) Review: actinomycetes diversity and biosynthetic capabilities of east coast of peninsular Malaysia coastal water and, 9) Climate change and coastal defenses in Malaysia: A review. The full-color figures have been included in this research book to better illustrate the features of some of the complex discussion parts. We strongly believe that this book is an added value to unveil the unexplored hidden treasures of the dynamic coastal ecosystem from Malaysia. We also foresee that the data presented in this book will act as a baseline to further explore research and improve management practices in the coastal ecosystem in Malaysia.

Editors
Akbar John
Zaima Azira Zainal Abidin
Ahmed Jalal Khan Chowdhury

FOREWORD

Malaysia is located in South East Asia comprising two regions namely, Peninsular Malaysia and the States of Sabah and Sarawak. The total land area covers 329,293 km^2 while the total coastline length is about 4,809 km. Additionally, there are about 1,000 islands and coral reefs belonging to Malaysia. The coastal zone is linked to both socio-economic and environmental significance. Majority of the populations occupy this area and it is also a center of economic activities encompassing aquaculture, oil and gas exploitation, agriculture, transportation and others. Mangrove areas are one of the most productive ecosystems on Earth. Mangroves as nursery and breeding ground for many fish and crustacean, and habitats for many wildlife species.

Progressive development in the coastal areas for urbanization and economic purposes has negatively impacted the environmental ecosystem. Thus, the need to establish sustainable development to ensure a balance between development and protection of the environment. Malaysia has expressed commitment to support and implement the 2030 Agenda and Sustainable Development Goals (SDGs) and sets out an ambitious plan of action for people, planet, prosperity, peace and partnership with the objective of leaving no one behind. Hence, implementation of sustainable development practices and holistic approaches in coastal areas is the key in achieving this objective.

I am delighted that the researchers from Kulliyyah of Science, IIUM prepared this book in its current form with the title on *'Bioprospects of coastal ecosystem and sustainable resource management'*. The book addressed various issues and bioprospecting potential of coastal ecosystems in a broader scale that open up opportunities for intellectual discussion in the near future. The advent of modern technology provides an insight on the coastal water potential and emphasized in this book. Therefore, I am optimistic that the findings in this publication will provide meaningful and impactful inputs to readers in upgrading their knowledge on coastal waters in Malaysia.

Prof. Dr. Kamaruzzaman Yunus
Campus Director
International Islamic University Malaysia,
Kuantan Campus
Pahang, Malaysia

FOREWORD

The holistic and integrated approaches for the sustainable development and utilization of coastal ecosystems is well discussed among the scientific community and policy makers and in recent years. In this regard, the importance of the ocean ecosystem and utilization of its resources is one of the main agenda of the United Nations sustainable development goal (SDG) in particular on SDG -14 'Live below water'. As the ocean covers a substantial portion of the earth's surface, it is estimated that over 3 billion people depend on marine and coastal resources for their livelihoods. Nowadays, the coastal ecosystem is increasingly degraded or destroyed by many human activities and eventually reduced its ability to support crucial ecosystem services. Eventually, the deterioration of the coastal ecosystem negatively impacted human well-being globally.

Having said this, the biological resources from the coastal ecosystem are less explored especially on the bioactive potential resource availability and its sustainable utilization. The present book on 'Bioprospecting coastal ecosystem towards sustainable resource management' is a timely effort by the researchers from International Islamic University Malaysia (IIUM) to compile the current threats impacting the coastal ecosystem management and exploring possible bioprospecting potential for human sustainable living. Considering the fact that Malaysia is one of the mega biodiversity nation and always prioritize the biodiversity as a key factor in research road map, I am confident that the scientific information shared by the researchers from Malaysia will act as a reference for further utilization of coastal resources in an effective manner and open up doors for the further research.

Although the book is primarily addressing the scientific findings, I observe the content and the intention of the editors and authors with the help of IIUM vision that insist to develop holistic individuals who can act as a 'Khalifa' (*ie.*, leader) and 'Rahmathal lil Alameen' (*ie.*, mercy to all the worlds) truly guided by the divine principles of '*Maqasid al-Shari'ah*'. I congratulate the contributors for their sincere and timely effort. In line with the vision and mission of the IIUM and the focus to achieve SDG 2030, I am confident that this book is an added value and informative to the broad-spectrum readers including academicians, researchers, policy makers, non-governmental organizations (NGOs) and students.

Prof. Dr. Ahmad Hafiz Bin Zulkifly
Deputy Rector (Responsible Research & Innovation)
International Islamic University Malaysia

iii

Seasonal Variations of Fish Diversity and Species Richness at the Coastal Water, Pekan, Pahang, Malaysia

Akbar John, B.[1*], Khuraisha, N.[2], Jalal, K.C.A[2*]. Najiah, M.[3] and Nadirah, M[3]

[1]*Institute of Oceanography and Maritime Studies (INOCEM),*
[2]*Department of Marine Science, Kulliyyah of Science, International Islamic University Malaysia (IIUM), Kuantan 25200, Pahang Malaysia.*
[3]*Faculty of Fisheries and Food Science, Universiti Malaysia Terengganu (UMT), 21030 Kuala Nerus, Terengganu*
Corresponding Author: akbarjohn50@gmail.com, jkchowdhury@iium.edu.my

ABSTRACT

This study was conducted from April 2019 to October 2019 to investigate the seasonal variations on fish diversity and species richness at coastal water Pekan, Pahang (Pantai Sepat, Cherok Paloh and Tanjung Selangor), Malaysia. A total of 5341 individual fishes were recorded which comprised 47 families and 108 species whereby 2444 individuals recorded during the non-monsoon season and 2897 individuals during the monsoon season. The most dominant families were Nemipteridae followed by Lutjanidae and Carangidae. The highest species richness was observed during the non-monsoon season with 95 species. Shannon-Weaver index (H'), Simpson's index of diversity (1-D), and Berger-Parker index were applied to demonstrate the species diversity, richness, evenness and dominance of fish in sampling areas and the overall values for non-monsoon season are 3.284, 0.9326 and 0.1335 respectively while for monsoon season are 2.766, 0.8798 and 0.2751 respectively. The high diversity index (Shannon-Weaver and Simpson) was observed in the non-monsoon season. This study has also demonstrated that the seasonal variations alone may not influence the number of species in a population along the coastal water Pekan. However, the status of fishing activities, collected fish species and water quality along the coastal water Pekan need to be monitored frequently for the sustainable harvesting of commercial species in coastal water of Pahang, Malaysia.

Keywords: Biodiversity; Fish Distribution; Ecology; Species Richness.

INTRODUCTION

Malaysia as one of the mega-biodiversity nation is the home for a total of 1951 species of freshwater and marine fishes belonging to 704 genera and 186 families of which half of the species are currently threatened and nearly one third of which are mostly from the marine and coral habitats (Chong et al., 2010). Specifically, the east coast of peninsular Malaysia is a susceptible fishing ground for by-catch activities from both Malaysian and Vietnamese fisherman. It was observed that indiscriminate fishing practices have been conducted along the coastal water Pahang for a decade, which could be responsible for gradually declining the fisheries resources in this fascinating coastal zone in the long run. In fact, the personal observation from the local fisherman also stated that the reducing number of the several species occurred due to several factors such as the massive intruders from the Vietnamese fishermen in international waters near Malaysian EEZ. Most of the species such as Starry triggerfish, sole fish, tiger shark and hammerhead shark are hard to find nowadays. According to Fazly et al., (2018), a foreign fishing boat found to be from Vietnam had intruded into Malaysian coastal waters to fish on 11th May 2019. In addition, Malaysian Society of Marine Sciences stated that the bauxite-contaminated red sea off the Pahang coastal area is bound to be a "dead sea": for up to three years. This is due to the increase in run-off from the ochre-red earth at the mines and the stockpiles located in Kuantan.

Fisheries management has always considered the relevant biological, technological, economic, social, environmental and commercial aspects of the industry towards ensuring effective conservation and management of all fishery resources. Determining the current resource potential have always been important considerations for fisheries managers. [DOF 2015]. Various management issues and challenges that have high impacts on fishing capacity are identified as below: i. Resources being overfished, ii. Inadequate updated data on fisheries resources, iii. Inadequate capacity and capability for monitoring and surveillance. vi. Insufficient public awareness and participation.

The unpublished studies conducted at Pantai Sepat by Jalal et al. (2012) showed that this area is not highly diversified with species. However, there were no previous studies on fish diversity along the coastal water Pekan, Pahang (Pantai Sepat to Tg. Selangor – the middle area of Kuala Pahang) which are the most vital area for fishing activities in the coastal water Pahang. Therefore, the present study was aimed to investigate the fish diversity and distribution and its seasonal variation in Pahang coastal water, Malaysia.

MATERIALS AND METHODS

Location of Fish Sampling
The study area is based on marine environments which extended along the coastal waters of Pahang, from 3.40155 ^{0}N to 3.34894 ^{0}N and 103.21174 ^{0}E to 103.25089 ^{0}E approximately 16 km (Fig 1). Coastal areas in Pahang such as Cherating, Teluk Cempedak, Tanjung Lumpur and Pantai Sepat are becoming the most attractive beaches by offering beautiful landscape and recreational activities (Azid et al., 2015; Tobergte & Curtis, 2013). Fish sampling was conducted from April 2019 to October 2019 by covering the fish diversity and distribution from Pantai Sepat, Cherok Paloh and Tanjung Selangor near Kuala Pahang during both monsoons and non-monsoon seasons. The sampling was conducted at noon as most of the fishermen landed their boat at this time. Five years (2014 to 2018) of accumulated data obtained from the World Weather Online showed that the highest wind speed occurred in 2016. The wettest month with the highest rainfall is December (563.9 mm) meanwhile the driest month with the lowest rainfall is February (142 mm) (MMD, 2019).

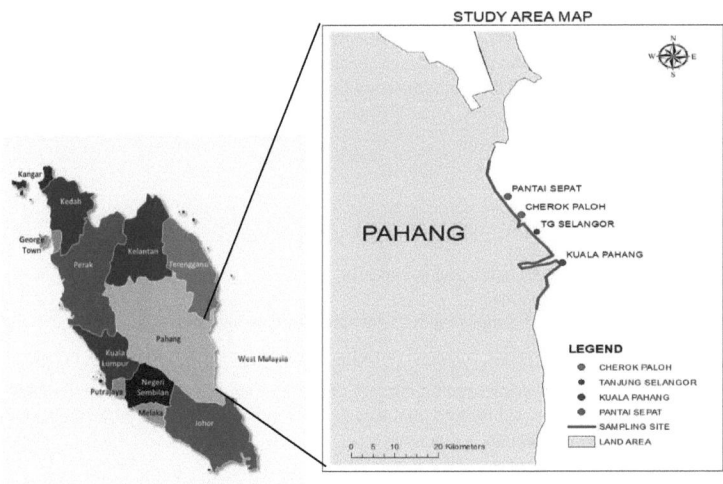

Fig. 1: Location of the Sampling Sites.

Data collection and fish identification
Specimens were collected from fish landing sites at the market near Pantai Sepat twice a month. The fishes were sorted into species and the standard lengths were taken by using a ruler and mounting board in the field where possible. All fish caught were counted and pictured using high resolution camera. Fish samples collected from the study areas were identified based on their morphometric and meristic characters according to technique as mentioned by Mansor et al, (1998); Ambak et al (2010). The environmental data such as temperature, and precipitation data were obtained from the World Weather Online.

Data and Software Analysis

Shannon Diversity Index
The diversity index calculated by using the Shannon-Weaver diversity index is used to characterize species diversity in a community and accounts for both abundance and evenness of the species present. This index is the most favored compared to the other indices. Usually, the values range between 0.0 – 5.0 and the results obtained fall between 1.5 – 3.5. Based on this index, the condition of the habitat can be identified. The structure of the habitat is considered stable and balanced when the values show above 3.5, meanwhile the values fall under 1.0 represent that the habitat structure is already degraded and polluted. Therefore, this index is very important to know the environment generally.

Formula
$$H' - \Sigma \left[(n_i / N) \times (\ln n_i / N) \right]$$

where,

H' : Shannon Diversity Index
n$_i$: Number of individuals belonging to i species
N : Total number of individuals

Simpson Diversity Index
Next, Simpson's dominance index (D) was used to quantify the biodiversity of habitat which considers the number of species, as well as the abundance of each species. This index varies between 0-1. However, the result is subtracted from 1 to correct the inverse proportion.

Formula
$$1 - D \left[\Sigma n_i (n_i - 1) \right] / N (N-1)$$

where,

D : Simpson Diversity Index
n$_i$: Number of individuals belonging to i species
N : Total number of individuals
Then, the reciprocal form (1/D) of Simpson index is adopted for data interpretation.

Berger- Parker Index
This index used to measure the proportional importance of the most abundant species. Same as Simpson index, the reciprocal of the index, 1/d is often used so that the increase in the value of the index represents an increase in diversity and a reduction in dominance.

Formula

$$d = N_{max} / N$$

where,

N_{max} : Number of individuals in the most abundant species

N : Total number of individuals in the sample

Species diversity and species richness indices Shannon-Weaver index (H'), Simpson index [1-D or 1/D], and Berger-Parker dominance Index were calculated using Biodiversity Pro V2 (Shannon and Weaver, 1949; Simpson, 1949; Caruso et al., 2007). All the software analysis is carried out using PAST326 meanwhile the statistical analysis is performed using SPSS 25v.

RESULTS

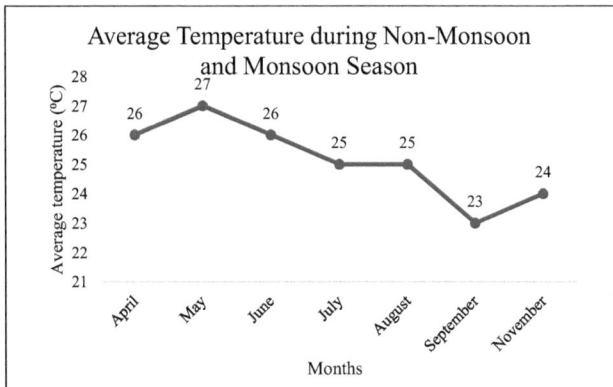

Fig. 2: Average temperature of Pekan, Pahang during non-monsoon and monsoon season

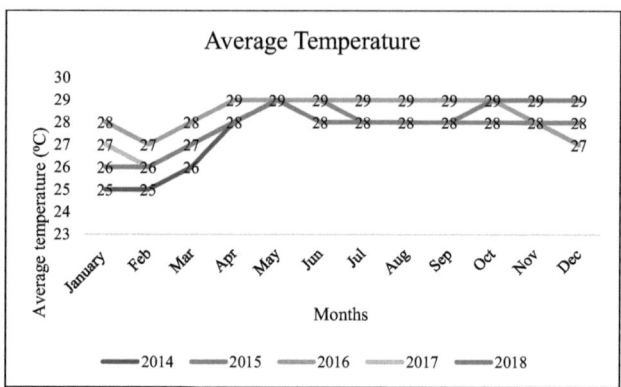

Fig. 3: Five years meteorological data of average temperature in Pekan, Pahang
(*sources: https://www.worldweatheronline.com/pekan-weather-history/pahang/my.aspx*)

The average temperature recorded during non-monsoon season ranged between 25°C and 27°C whereby the lowest was recorded in July and August and the highest was recorded in May (Fig 2). During the monsoon season, the highest average temperature was recorded in October (24°C) and the lowest (23°C) was recorded in September. The five years of meteorological data (2014-2018) found that the temperature varied between 25°C and 29°C (Fig 3). The temperature is slightly increased by 1°C each year. From July to August, the temperature is constantly stagnant at 28°C from 2014 to 2018.

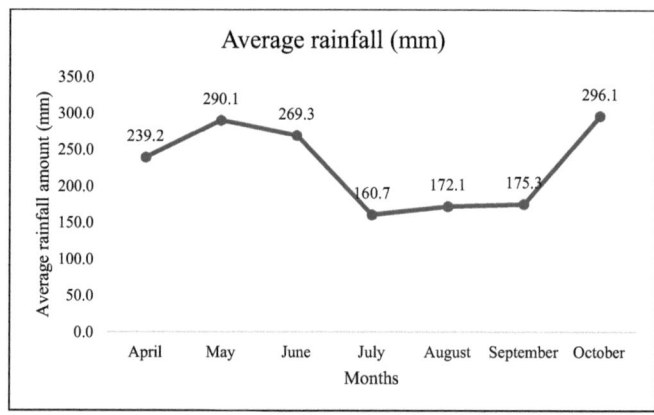

Fig. 4: Average rainfall amount (mm) of Pekan, Pahang during non-monsoon and monsoon

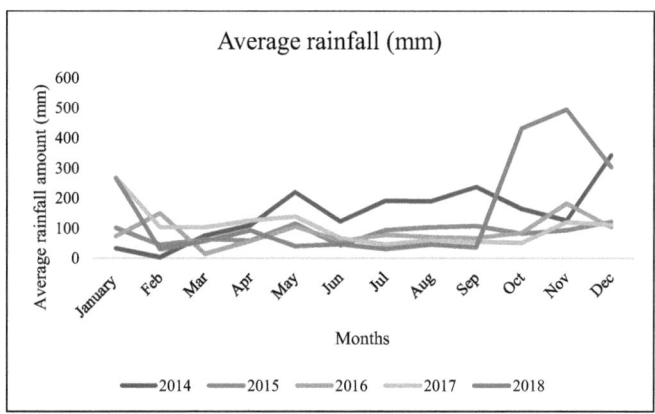

Fig. 5: Five years data of average rainfall amount (mm) at Pekan, Pahang
(*sources: https://www.worldweatheronline.com/pekan-weather-history/pahang/my.aspx*)

During the non-monsoon season, the highest average rainfall amount (mm) was recorded in May (290.1 mm) and the lowest was recorded in July (160.7 mm). Meanwhile, during monsoon season, the highest average rainfall amount (mm) was recorded in October (296.1 mm) and the lowest was recorded in September (175.3 mm). Five years' trend of meteorological data showed that the rainfall occurred at maximum level 494.1 mm during November at Pekan, Pahang where the minimum level found 2.53 mm during February (Fig. 4). Air temperature varied between 25°C and 29°C along the years from 2014 until 2018 (Fig 5).

Table 1: List of species identified in the coastal water Pekan, Pahang

Class	Order	Family	Species
Actinopterygii	Beryciformes	Holocentridae	*Sargocentron rubrum*
	Beryciformes	Holocentridae	*Myripistis hexagona*
	Mugiliformes	Mugilidae	*Valamugil speigelri*
	Clupeiformes	Clupeidae	*Sardinella melanura*
	Clupeiformes	Chirocentridae	*Chirocentrus dorab*
	Clupeiformes	Eugraulidae	*Thryssa mystax*
	Siluriformes	Ariidae	*Arius maculatus*
	Siluriformes	Plotosidae	*Plotosus canius*
	Gadiformes	Batrachoididae	*Batrachomoeus trispinosus*
	Perciformes	Carangidae	*Selaroides leptolepis*

Perciformes	Carangidae	*Selar boops*
Perciformes	Carangidae	*Atule mate*
Perciformes	Carangidae	*Tranchinotus blochii*
Perciformes	Carangidae	*Alectis indicus*
Perciformes	Carangidae	*Alectis ciliaris*
Perciformes	Carangidae	*Carangoides malabaricus*
Perciformes	Carangidae	*Megalaspis cordyla*
Perciformes	Caesionidae	*Caesio cunning*
Perciformes	Caesionidae	*Caesio caerulaurea*
Perciformes	Chaetodontidae	*Coradion chrysozonus*
Perciformes	Chaetodontidae	*Chelmon rostratus*
Perciformes	Drepaneidae	*Drepane longimana*
Perciformes	Drepaneidae	*Drepane punctata*
Perciformes	Ephippidae	*Platax teira*
Perciformes	Gerreidae	*Gerres oyena*
Perciformes	Gerreidae	*Gerres erythrourus*
Perciformes	Haemulidae	*Pomadasys maculatus*
Perciformes	Haemulidae	*Pomadasys kaakan*
Perciformes	Haemulidae	*Diagramma punctatum*
Perciformes	Haemulidae	*Plectorhincus gaterinus*
Perciformes	Lactariidae	*Lactarius lactarius*
Perciformes	Lethrinidae	*Lethrinus lentjan*
Perciformes	Lethrinidae	*Letrinus miniatus*
Perciformes	Lethrinidae	*Lethrinus genivittatus*
Perciformes	Lethrinidae	*Letrinus ornatus*
Perciformes	Lethrinidae	*Gymnocranius frenatus*
Perciformes	Lutjanidae	*Lutjanus vitta*
Perciformes	Lutjanidae	*Lutjanus ruselli*
Perciformes	Lutjanidae	*Lutjanus malabaricus*

Perciformes	Lutjanidae	*Lutjanus lutjanus*
Perciformes	Mullidae	*Upenus tragula*
Perciformes	Mullidae	*Upeneus japonicus*
Perciformes	Nemipteridae	*Pentapodus setosus*
Perciformes	Nemipteridae	*Scolopsis monograma*
Perciformes	Nemipteridae	*Nemipterus furcosus*
Perciformes	Nemipteridae	*Scolopsis taenioptera*
Perciformes	Nemipteridae	*Scolopis affinis*
Perciformes	Pomacanthidae	*Chaetodontoplus mesoleucus*
Perciformes	Rachycentridae	*Rachycentron canadum*
Perciformes	Serranidae	*Epinephelus areolatus*
Perciformes	Serranidae	*Cephalopholis urodeta*
Perciformes	Serranidae	*Cephalopholis cyanostigma*
Perciformes	Serranidae	*Epinephelus formosa*
Perciformes	Serranidae	*Epinephelus coiodes*
Perciformes	Serranidae	*Cephalopholis boenack*
Perciformes	Serranidae	*Plectropomus maculatus*
Perciformes	Serranidae	*Diplorion bifasciatum*
Perciformes	Serranidae	*Epinephelus sexfasciatus*
Perciformes	Labridae	*Choerodon schoenleinii*
Perciformes	Labridae	*Cheilinus trilobatus*
Perciformes	Labridae	*Cheilinus chlorourus*
Perciformes	Polynemidae	*Eleutheronema tetradactylus*
Perciformes	Pomacentridae	*Abudefduf bengalensis*
Perciformes	Pomacentridae	*Pomacanthus annularis*
Perciformes	Scaridae	*Scarus ghobban*
Perciformes	Scatophagidae	*Siganus guttatus*
Perciformes	Scombridae	*Scomberoides commersonnianus*
Perciformes	Scombridae	*Scomberoides tala*

	Perciformes	Scombridae	*Rastrelliger brachysoma*
	Perciformes	Scombridae	*Rastrelliger kanagurta*
	Perciformes	Sciaenidae	*Paranibea semiluctuosa*
	Perciformes	Sparidae	*Terapon jarbua*
	Perciformes	Sparidae	*Dextex tumifrons*
	Perciformes	Sphyreanidae	*Sphyraena flavicaudas*
	Perciformes	Sphyreanidae	*Sphyraena putnamae*
	Perciformes	Sphyreanidae	*Sphyraena forsteri*
	Perciformes	Sphyreanidae	*Sphyraena jello*
	Perciformes	Siganidae	*Siganus javus*
	Perciformes	Siganidae	*Siganus fuscescens*
	Perciformes	Siganidae	*Siganus vulpinus*
	Perciformes	Siganidae	*Siganus canaliculatus*
	Perciformes	Toxotidae	*Toxotes chatareus*
	Pleuronectiformes	Cynoglossidae	*Cynoglossus bilineatus*
	Pleuronectiformes	Psettodidae	*Psettodes erumei*
	Clupeiformes	Clupeidae	*Sardinella melanura*
	Carcharhiniformes	Scyliorhinidae	*Atelomycterus marmoratus*
	Orectolobiformes	Hemiscyllidae	*Chiloscyllium griseum*
	Orectolobiformes	Hemiscyllidae	*Chiloscyllium punctatum*
	Orectolobiformes	Brachaeluridae	*Brachaelurus colcloughi*
	Myliobatiformes	Dasyatidae	*Taeniura lymma*
	Myliobatiformes	Dasyatidae	*Dasyatis ushie*
Chondrichthyes	Myliobatiformes	Dasyatidae	*Pastinachus sephen*
	Myliobatiformes	Dasyatidae	*Himantura gerradi*
	Myliobatiformes	Dasyatidae	*Dasyatis parvonigra*
	Myliobatiformes	Myliobatidae	*Aetobatus narinari*
	Rajiformes	Rajidae	*Rhycobatus australiae*
	Tetraodontiformes	Balistiidae	*Abalistes stellaris*

Tetraodontiformes	Diodontidae	*Diodon hystix*
Tetraodontiformes	Monocanthidae	*Chaetodermis penicilligerus*
Tetraodontiformes	Monocanthidae	*Monacanthus chinensis*
Tetraodontiformes	Monacanthidae	*Aluterus scriptus*
Tetraodontiformes	Monocanthidae	*Aluterus monocerus*
Tetraodontiformes	Monocanthidae	*Pseudomonacanthus macrurus*
Tetraodontiformes	Ostraciidae	*Ostracion cubicus*
Tetraodontiformes	Ostraciidae	*Ostracion nasus*
Tetraodontiformes	Tetraodontidae	*Lagocephalus suezensis*
Tetraodontiformes	Tetraodontidae	*Arothron immaculatus*
Tetraodontiformes	Tetraodontidae	*Arothron mappa*

A total of 5341 individuals were recorded which comprises 47 families belonging to 75 genera of 108 species throughout the sampling period (April 2019 until October 2019) from the coastal water Pekan, Pahang from (Table 1). The captured fish was dominated by Nemipteridae followed by Lutjanidae and Carangidae family. These 47 families were categorized under the class of Chondrichthyes and Osteichthyes, which played a vital role to make the species composition of fishes in coastal water Pekan. Class Osteichthyes (ray-finned fishes) was observed as the largest class of vertebrates along with 50 species found in this study. Those fishes in this class were identified with fin rays and scales on their body (ganoid, cycloid or ctenoid).

Among the other families in this study, the Nemipteridae family was dominant by contributing 36.01% of the total fish caught in the study area during non-monsoon and monsoon season with the diversity index (H') of 1.376 and 1.115 respectively. The Nemipteridae family consist of 5 species which is; *Pentapodus setosus, Scolopsis monogramma, Nemipterus furcosus, Scolopsis taenioptera* and *Scolopsis affinis*. This family or also known as threadfin bream are a common demersal fish of the Indo-Pacific which comprises 3 genera namely; *Nemipterus, Pentapodus* and *Scolopsis*. Among all species in the family Nemipteridae, *Nemipterus furcosus* was the dominant among all 5 species.

Based on samples collected, Nemipterus *furcosus* is considered as the dominant species due to the highest abundance whereby it contributes 43% to the number of individuals caught from the sampling area. The highest number recorded was in October. Second highest species were also from family Nemipteridae, which is *Pentapodus setosus* contributing 29%. *Scolopsis monogramma, Scolopsis taenioptera* and *Scolopsis affinis* was recorded by the overall number of individuals caught 413, 159 and 66 respectively. *Nemipterus furcosus* and *Scolopsis taenioptera* was caught highest in October with 542 individuals and 80 individuals, *Scolopsis monogramma,* was recorded highest in August, the same as *Scolopsis affinis*.

Table 2: seasonal variation in the Percentage abundance of fishes (%) from the Coastal Water Pekan, Pahang

Non monsoon		Monsoon	
Family	Abundance (%)	Families	Abundance (%)
Nemipteridae	36.01%	Nemipteridae	48.71%
Lutjanidae	21.85%	Lutjanidae	15.91%
Carangidae	5.73%	Carangidae	11.25%
Serranidae	3.89%	Serranidae	4.45%
Sparidae	3.31%	Haemulidae	3.59%
Siganidae	2.70%	Siganidae	2.52%
Mullidae	2.54%	Monocanthidae	2.38%
Caesionidae	2.25%	Sparidae	1.79%
Dasyatidae	2.25%	Scombridae	1.59%
Haemulidae	1.55%	Tetraodontidae	1.24%
Monocanthidae	1.55%	Caesionidae	1.24%
Rajidae	1.51%	Mullidae	0.90%
Ariidae	1.31%	Ariidae	0.86%
Scaridae	1.19%	Lethrinidae	0.76%
Chirocentridae	1.15%	Holocentridae	0.48%
Tetraodontidae	1.10%	Sphyreanidae	0.48%
Hemiscyllidae	1.06%	Gerreidae	0.35%
Brachaeluridae	0.90%	Scaridae	0.24%
Scombridae	0.90%	Brachaeluridae	0.17%
Sciaenidae	0.86%	Eugraulidae	0.14%
Gerreidae	0.82%	Ostraciidae	0.14%
Holocentridae	0.82%	Balistiidae	0.10%
Scatophagidae	0.74%	Chaetodontidae	0.10%
Sphyreanidae	0.74%	Cyglossidae	0.07%
Lethrinidae	0.41%	Drepaneidae	0.07%
Drepaneidae	0.37%	Ephippidae	0.07%

Chaetodontidae	0.33%	Batrachoididae	0.03%
Toxotidae	0.33%	Chirocentridae	0.03%
Polynemidae	0.49%	Dasyatidae	0.03%
Ostraciidae	0.29%	Hemiscyllidae	0.10%
Labridae	0.20%	Lactariidae	0.10%
Scyliorhinidae	0.20%	Labridae	0.03%
Pomacentridae	0.08%	Pomacentridae	0.03%
Mugilidae	0.08%		
Ephippidae	0.08%		
Eugraulidae	0.08%		
Rachycentridae	0.08%		
Clupeidae	0.04%		
Diodontidae	0.04%		
Myliobatidae	0.04%		
Pomacanthidae	0.04%		
Psettodidae	0.04%		
Plotosidae	0.04%		

The Nemipteridae family is a bottom-living fish which lives in mud and sand bottoms in coastal inshore as well as offshore shelf waters. The characters of this family are elongated to moderately deep, compressed, small to medium sized sparoid fishes. In *Nemipterus* and *Pentapodus,* the mouth is terminal, small to moderate; moderately protrusible; teeth in jaws conical, enlarged canines present. The body colour appeared to be extremely emerging, often pinkish or reddish with red, yellow or blue markings. Fishermen often catch these threadfin bream since it has a high demand in the market.

Family Lutjanidae was the second highest family caught in this study area by contributing 21.85% of all fish caught during the non-monsoon season. The family of Lutjanidae from the sampling area consisted of *Lutjanus vitta, Lutjanus ruselli,* and *Lutjanus lutjanus.* The percentage of species contributed from this family were; *Lutjanus vitta*: 38%, *Lutjanus ruselli*: 1%, *Lutjanus lutjanus*: 61% (Table 2).

Table 3: The diversity and dominance index of the fishes identified from the sampling locations.

Seasonal variations	Total No. of species found	H'	1-D	BP
Non-monsoon	92	3.284	0.9326	0.1335
Monsoon	67	2.766	0.8798	0.2751

The value of the Shannon Weaver diversity index (H'), Simpson Index and Berger Parker Index were calculated according to seasonal variations. After calculating whole samples (108), total H' value was found 3.288 during non-monsoon season and 2.766 during monsoon season. There is no significant difference (p>0.05) between two monsoons. During the non-monsoon season, the highest Shannon diversity index (2.978) was found in June and lowest (2.466) was found in May. Meanwhile, the highest Shannon diversity index (2.884) was found in September and the lowest (2.244) was found in October during monsoon season. The Simpson diversity index, (1/D) was highest (0.9327) during non-monsoon season compared to non-monsoon season (0.8798). Berger Parker Dominance (a/d) index showed that species dominance was higher during monsoon season with 0.2751 compared to non-monsoon season (0.1334) (Table 3)

DISCUSSION
The declining fishes commonly occurred due to several factors such as overexploitation of species, the introduction of invasive species, pollution from urban, industrial as well as habitat loss of aquatic biodiversity in both freshwater and marine environments. As a result, valuable aquatic resources are becoming increasingly prone to both natural and artificial environmental changes. Thus, a conservation strategy to protect and conserve aquatic life is necessary to maintain the balance of nature and support the availability of resources for future generations (Ahmad Azfar, 2009). The South China Sea lies in the tropical zone of the western Pacific Ocean, off the southeast corner of the Asian continent, and is known for both its high productivity and the rich diversity of plants and animals. In this study, a total of 5341 individuals were recorded which composes of 47 families and 108 species from the coastal water Pekan, Pahang whereby 2444 individuals recorded during the non-monsoon season and 2897 individuals during the monsoon season.

Similar studies have been conducted by other researchers in the South China Sea. Randall and Lim (2000) listed at least 3,365 species of marine fishes from the South China Sea. Mohsin and Ambak (1996) reported 710 species of marine fishes from Malaysian waters and adjacent seas. Adrim et al. (2004) recorded 430 marine fish species from the Anambas and Natuna islands on the Sunda Shelf between the Malay Peninsula and Borneo in the South China Sea. More recently, Ambak et al. (2010) estimated 2,243 fish species occurring in Malaysian waters and 26% of over 441 fish species recorded by Matsunuma et al. (2011) in Terengganu water.

The field surveys of fishes in Terengganu in 2008-2009 recorded 441 marine and estuarine fish species of 108 families whereby make up around 13% of over 3,365 fish species recorded by Randall and Lim (2000) from the Southchina Sea. The morphology, ecology, distribution, specimens with photos, and literature of fishes (300 families with 3086 species) that mainly found in the Southchina Sea were collected by the *Fish Database of Taiwan* (Shao 2011).

According to Wang et al., (2012), there were 95 species in 86 genera of 69 families were identified by using DNA Barcoding from two regions in Southchina Sea; Spratly Islands and Beibu Gulf. Also, Adrim et al., (2004) recorded 430 marine fish species from the Anambas and Natuna islands on the Sunda Shelf between the Malay Peninsula and Borneo in the South China Sea. Mohsin and Ambak (1996) reported 710 species of marine fishes from Malaysian waters and adjacent seas.

Based on the Shannon-Weaver index, the non-monsoon season is more diverse compared to the monsoon season. However, there is no significant difference between the two seasons. Also, the Simpson diversity index (1/d) showed that the non-monsoon season is more diverse than the monsoon season. According to Alonso et al., (2017), the yearly monsoon cycle is a major natural force which influences marine organisms in tropical regions. A study was conducted by (Al, 2007) reported that temperature significantly affects marine larval dispersal due to the rate of biochemical processes in organisms controlled by temperature. As a result, the population, species and community-level processes were affected. By the fluctuation in temperature, the number and diversity of adult species are changing in the marine environment as the larval development time is changing. It was evident that the values of water quality parameters or the effect of increasing fishing pressure would be responsible for differences in species diversity in various habitats of the sea (Komsari et al, 2015; Jalal, et al, 2012 a, b). Our water quality data from the Meteorological Department along the coastal water Pekan have shown that there were no major fluctuations in physical parameters (Temperature and Rainfall data) during the study period. Perhaps the amount of rainfall with the existing range of temperature could be two principal factors in triggering the captured fishes to initiate spawning activities and increasing in abundance of three families (Nemipteridae followed by Lutjanidae and Carangidae family) in the sampling area.

In this study, the Shannon-Weaver Nemipteridae family recorded the highest Shannon-Weaver index in the monsoon season compared to the non-monsoon season. This area could be a spawning area as it was reported by the fisherman when they observed fish, eggs and fry around the study area. Besides, the fishes belonging to this family can move mostly in the form of schooling to feed mainly on other small fishes, cephalopods, crustaceans and polychaetes. In fact, the highest catch of this family also might be due to the high demand in the market as they are commercial and artisanal fisheries. Similarly, the families found in this area are Lutjanidae, Caesionidae, Lethrinidae and Haemulidae. It was observed that different species have different spawning time and habitat.

Therefore, the second highest individuals caught along the sampling period are Lutjanidae. This family is also known as snappers and contains more than 100 species of tropical and subtropical fish. The Shannon-Weaver index of this fish showed higher in monsoon season compared to non-monsoon season. According to Pacific Community, this family commonly spawns along the years in warmer waters, but, during the warmer months, they travel to cooler waters particularly along outer reefs and channels to breed. According to Al (2007), the distance larvae travelled varied with ocean temperature. It was found out that the larvae from the same species travel more in colder waters compared to warmer waters. The fish fry in cold waters develop more slowly and drift further before starting their next development stage since the sluggish metabolisms caused by the cold temperatures. The fertilized eggs in most reef-related snappers hatch that drift with the currents for about one-month hatch into small forms. After 3 to 8 years, the juveniles become a mature adult and exposed to open coastal water areas. Thus, they are easily caught as they gather in large groups to breed which was evident during our study period along the fishing areas of the coastal water Pekan.

The third highly diverse family recorded for this study was Carangidae which contributed 5.73% during the non-monsoon season and 11.25% during monsoon season. The favourable habitat of this family is the coastal water in tropical and temperate waters around the world. Mostly the species move in schools except for *Alectis*; some species largely distributed and the young usually can be found in brackish environments, others (*Elagatis* and *Naucrates*) are pelagic fish which commonly found at or near the surface in oceanic waters. Among these families, there were several species identified; *Selaroides leptolepis, Selar boops, Atule mate, Tranchinotus blochii, Alectis indicus, Alectis ciliaris, Carangoides malabaricus,* and *Megalaspis cordyla. Atule mate* is the highest individual caught from both monsoon and non-monsoon. According to Mundy (2005), the adults can be found in mangrove areas and coastal bays in pelagic waters. In addition, a form of school can be recorded in inshore waters (Smith-Vaniz., 1999). Their food predominantly on crustaceans and planktonic vertebrates such as copepods (Allen et al., 2012; Fischer et al., 1990).

CONCLUSION
A total of 5341 individuals comprising 75 genera, 47 families and 108 species was recorded in the coastal water Pekan, Pahang, Malaysia. The captured fish was dominated by Nemipteridae followed by Lutjanidae and the Carangidae family was highly diverse in the study area. The presence of different size fry in the fishing net indicated that the spawning area of the species of these three (3) families might be located along the coastal water Pekan. Overall, the high species diversity in the sampling area could suggest that there

could be many successful species and a more stable ecosystem. Furthermore, a complex food web and environmental changes are less likely to be damaging to the ecosystem in the vicinity of coastal water of Pekan.

Nevertheless, the fishing activities along the coastal water need to be controlled towards the discriminant way for the sustainable development of these valuable commercial species in this fascinating coastal water of Pekan, Pahang, Malaysia. Monitoring programs for fisheries should involve periodic sampling using techniques such as experimental fishing and aerial surveying of fishermen in order to determine species diversity and socioeconomics of the fish community. The information obtained could then be used to determine the healthiness of the coastal water, estuarine and river system as well to initiate the suitable management and conservation programs along the South China Sea.

REFERENCES

Adrim, M., I.-S. Chen, Z.-P. Chen, K. K. P. Lim, H. H. Tan, Y. Yusof, and Z. Jaafar. (2004). Marine fishes recorded from the Anambas and Natuna Islands, South China Sea. Raffles Bull. Zool. Suppl., (11): 117–130.

Ahmad Azfar, M. (2009) Diversity and Distribution of Fishes in Pahang Estuary, Malaysia. MS thesis. 196 pp.

Al, M. I. O. et. (2007). How do Changes in Ocean Temperature affect Marine Ecosystems?, (52), 2007–2007. From http://ec.europa.eu/environment/integration/research/newsalert/pdf/52na2.pdf

Allen, G.R. and M.V. Erdmann, 2012. Reef fishes of the East Indies. Perth, Australia: University of Hawai'i Press, Volumes I-III. Tropical Reef Research.

Alonso Aller, E., Jiddawi, N. S., & Eklöf, J. S. (2017). Marine protected areas increase temporal stability of community structure, but not density or diversity, of tropical seagrass fish communities. *PLoS ONE, 12*(8), 1–23. https://doi.org/10.1371/journal.pone.0183999

Ambak, M.A., Mansor, M.I., Zaidi, M.Z. and Mazlan, A. G (2010). *Fishes in Malaysia*. 315 pp.

Azid, A., Noraini, C., Hasnam, C., Juahir, H., Amran, M.A., Toriman, M.E. & Kamarudin, A. 2015. Coastal erosion measurement along Tanjung Lumpur to Cherok Paloh, Pahang during the Northeast Monsoon Season. *Journal Teknologi* 1: 27-34.

Caruso, T., Pigino, G., Bernini, F., Bargagli, R., & Migliorini, M. (2007). The Berger– Parker index as an effective tool for monitoring the biodiversity of disturbed soils: a case study on Mediterranean oribatid (Acari: Oribatida) assemblages. *Biodiversity and Conservation, 16*(12), 3277-3285.

Chong, V. C., Jamizan, A. R., Yazid, Z., Rizman, I., Ali, S. H. & Natin, P. (2010). Diversity and abundance of fish and invertebrates of Semerak estuary and adjacent inshore waters, Kelantan. *Malaysian Journal of Science* **29,** 91–106.

Department of Fisheries (2015) National plan for action for the management of fishing capacity in Malaysia (Plan 2). 50 pp.

Fazly Amri Mohd, Khairul Nizam Abdul Maulud, Rawshan Ara Begum, Siti Norsakinah Selamat, & Othman A.Karim. (2018). Impact of Shoreline Changes to Pahang Coastal Area by Using Geospatial Technology. *Sains Malaysiana, 47*(5), 991–997.

Fischer, W., I. Sousa, C. Silva, A. de Freitas, J.M. Poutiers, W. Schneider, T.C. Borges, J.P. Feral and A. Massinga, 1990. Fichas FAO de identificaçao de espécies para actividades de pesca. Guia de campo das espécies comerciais marinhas e de águas salobras de Moçambique. Publicaçao preparada em collaboraçao com o Instituto de Investigaçao Pesquiera de Moçambique, com financiamento do Projecto PNUD/FAO MOZ/86/030 e de NORAD. Roma, FAO. 1990. 424 p.

Jalal, K.C.A, Kamaruzzaman, Y. Arshad A., Arafatur, R., Rahman, M. F. (2012 a). Diversity and distribution of fishes in tropical estuary Kuantan, Pahang, Malaysia. Pakistan Journal of Biological Sciences, 15 (12), pp. 576-582.

Jalal, K.C.A, M. Ahmad Azfar, B. Akbar John, Y.B. Kamaruzzaman and S. Shahbudin. (2012 b). Diversity and Community Composition of Fishes in Tropical Estuary Pahang Malaysia. Pakistan Journal of Zoology. 44(1), 181-187.

Komsari, M.S., Barni, A., Khara, H. (2015) Growth and population on the structure of the European Perch *Percafluviatilis Linnaeus*, 1758 (Osteichthyes: Percidae) in the Anzali wetland south-west Caspian Sea. Ind, J. Fish. 62(1):6-11.

Mansor, M.I., Kohno, H., Ida, H., Nakamura, H. T., Aznan, Z. & Abdullah, S. (eds.), (1998). Field Guide to important commercial marine fishes of the South China Sea. SEAFDEC/MFRDMD/SP/2.

Matsunuma, M., Motomura, H., Matsuura, K., Shazili, N. A. M., & Ambak, M. A. (2011). *Fishes of Terengganu East coast of Malay Peninsula, Malaysia. National Museum of Nature and Science.* Retrieved from http://www.museum.kagoshima-u.ac.jp/staff/motomura/TFG_lowres.pdf

MMD. (2011). Malaysian Meteorological Department monthly rainfall review. (2011). From: http://www.met.gov.my/?lang=en

Mohsin, A. K. M. and M. A. Ambak. 1996. Marine fishes and fisheries of Malaysia and neighbouring countries. Universiti Pertanian Press, Serdang, iv + xxxvi + 744 pp.

Mundy B.C., (2005). Checklist of the fishes of the Hawaiian Archipelago. Bishop Mus. Bull. Zool. (6):1-704

Randall J.E., Lim KKP, Alien GR, Amaoka K, Anderson WD, Jr., Bellwood DR, Bohlke EB, Bradbury MG, Carpenter KE, Caruso JH, Cohen AC, Cohen DM. (2000). A checklist of the fishes of the South China Sea. Raffles Bull Zool supplement: 569–667.

Shannon, C. E., and Weaver, W., 1949. *The Mathematical Theory of Communication.*

Shao K.T., (2011). The Fish Datebase of Taiwan. WWW Web electronic publication. version 2009/1.

Simpson, E. H. (1949). Measurement of diversity. *Nature 163,* 688

Smith-Vaniz, W.F., 1999. Carangidae. Jacks and scads (also trevallies, queenfishes, runners, amberjacks, pilotfishes, pampanos, etc.). p. 2659-2756. In K.E. Carpenter and V.H. Niem (eds.) FAO species identification guide for fishery purposes. The living marine resources of the Western Central Pacific. Vol. 4. Bony fishes part 2 (Mugilidae to Carangidae). Rome, FAO. 2069-2790 p.

Tobergte, D.R. & Curtis, S. 2013. Malaysia east coast region. *Journal of Chemical* Urbana: University of Illinois Press.

Wang, Z. D., Guo, Y. S., Liu, X. M., Fan, Y. B., & Liu, C. W. (2012). DNA barcoding South China Sea fishes. *Mitochondrial DNA, 23*(5), 405–410. https://doi.org/10.3109/19401736.2012.710204

Study of Glucose-6-Phosphate Dehydrogenase Activity Assay in Mangrove Streptomyces for Actinohordin and Undercylprodigiosin Production

Azizan, N.H. *[1], Zainal Abidin, Z.A.[1], Sharif, M.F.[1] and Mohd Maizam, A.F.[1]

[1]*Department of Biotechnology, Kulliyyah of Science, International Islamic University Malaysia, Jalan Sultan Ahmad Shah, Bandar Indera Mahkota, 25200, Kuantan, Pahang, Malaysia.*
Corresponding author: fizahazizan@iium.edu.my

ABSTRACT

This study evaluates the potential of using glucose-6-phosphate dehydrogenase activity assay for Actinohordin and Undecylprodigiosin productions from mangrove Streptomyces. Previously, there were several methods used to screen antimicrobial activities such as agar spot test and disc diffusion assay, but those are lengthy screening methods and time consuming. Thus, to overcome the limitations plate-based assay is suggested to enable rapid screening on secondary metabolite production of numerous samples at one time. The development of plate-based assay was performed by optimizing glucose-6-phosphate dehydrogenase activity assay. This coupled assay was based on the production of dihydronicotinamide-adenine dinucleotide phosphate (NADPH) whereby a right combination of nicotinamide adenine dinucleotide phosphate (NADP) and glucose-6-phosphate (G6P) were refined. The production of NADPH was measured at absorbance of 340 nm where reduced cofactor NADPH are absorbed readily at this wavelength. Sample with different concentrations of crude lysate was subjected to various substrates concentration to obtain the best activity curve. Even though elucidating clear patterns is speculative, it is believed that some improvements or optimizations of this study could offer promising knowledge which can serve as useful reference in future.

Keywords: *Actinohordin, Dihydronicotinamide-Adenine Dinucleotide Phosphate, Nicotinamide Adenine Dinucleotide and Undecylprodigiosin.*

INTRODUCTION

Actinomycetes are gram-positive filamentous bacteria that produce aerial hypae and differentiate into chains of spores (Kämpfer, 2015; Barka *et. al.*, 2016). They can be found in soil, freshwater and marine environments. They produced various useful compounds known as secondary metabolites with important applications such as antibiotics tetracycline, erythromycin, vancomycin and streptomycin (Weber *et al,* 2015). During the past thirty years, researchers have shown an increased interest towards antibiotics-producing bacteria as they give many benefits in human medicine as well as in commercial production.

Previously, the antimicrobial activities of secondary metabolites were assessed either by covering an isolation plate with indicator organism or agar-spot test where it has been used to detect antagonistic activity between bacteria (Kun, 2003). However, these methods have major limitations where potential contamination of selected colonies with indicator organisms could occur. In addition, they are lengthy screening methods as only one indicator organism can be applied to each isolation plate at a time. Apart from that, HPLC is also one of the options of screening methods, yet time consuming (Ethiraj *et al.,* 2011).

Nonetheless, secondary metabolites are typically produced at a very low amount in nature. Thus, many researches have been done previously to study the metabolic network of central carbon metabolism, precursors and cofactors required in synthesizing secondary metabolites to improve the product yield (Fan *et al.,* 2016). It is found that the amounts of precursors for secondary metabolite production required from primary metabolism gradually becomes limited as the product yield increases. Therefore, it is necessary to

17

supply an adequate number of precursors which is generally provided by catabolism of carbon substrates to obtain high yield of secondary metabolites.

Thus, to optimize the enzyme assay, a study was designed to induce the production of two secondary metabolic compounds, actinohordin (ACT) and undecylprodigiosin (RED) by targeting the pentose phosphate pathway (PPP) of *Streptomyces*. This is performed by promoting the conversion of the first enzyme of the pathway, which is glucose-6-phosphate dehydrogenase (G6PDH) by finding the best ratio combination of its substrates; glucose-6-phosphate (G6P) and nicotinamide adenine dinucleotide (NAD). This is to ensure G6PDH enzymes are supplied with adequate amounts of substrate in order to maximize the production of NADPH prior catalyzing the second metabolic pathway which in concert will enhance the antibiotic production as suggested by Gunarson *et al.,* (2004). Essentially, NADPH is the reducing agent used in the process of making secondary metabolites.

ACTINOMYCETES
The name actinomycetes was derived from Greek word "aktis" which means a ray and "mykes" which refers to fungus. This name was given by looking at their morphology where they possess characteristics of both bacteria and fungi (Das *et al.,* 2008) but yet, they are categorized into bacteria kingdom (Madigan *et al.,* 2009). They contain DNA rich in G+C at about 57-75% (Lo *et al.,* 2002) which are phylogenetically related from evidence of 16s ribosomal cataloguing and DNA: rRNA pairing studies by Goodfellow & Williams (1983). They are characterized by a complex life cycle, as described by phylum Actinobacteria, which represents one of the largest taxonomic units among the 18 major lineages currently recognized within the Domain Bacteria (Ventura *et al.,* 2007).

Actinomycetes are commonly found in both terrestrial and aquatic ecosystems which mainly in soil. They play an important role in recycling refractory biomaterials by decomposing complex mixtures of polymers in dead plants, animals and fungal materials resulting in production of many extracellular enzymes which are conductive to crop production (Chaudhary *et al.,* 2013). In addition, actinomycetes also give major effects in biological buffering of soils, biological control of environments by nitrogen fixation and degradation of high molecular weight compounds like hydrocarbons in the polluted soil. Thus, these microorganisms play vital roles in maintaining our ecosystems.

Above all, actinomycetes are valuable bacteria which are commonly known due to their ability to produce secondary metabolites. Berdy (2005) reported that 10000 out of 23000 bioactive secondary metabolites produced by microorganisms originate from actinomycetes bacteria, representing 45 % of all bioactive microbes discovered. Among various genera of actinomycetes, the major producers of commercially bioactive compounds are *Streptomyces, Saccharopolyspora, Amycolatopsis, Micromonospora and Actinoplanes* (Solanki *et al.,* 2008).

Streptomycetes coelicolor A3 (2)
Streptomycetes species are aerobic and gram-positive bacteria that show filamentous growth from a single spore. A network of branched filaments called as a substrate mycelium will be formed when their filaments grow through tip extension and branching (Dyson, 2011). They are widely recognized as they are the major producer and have produced a total of 7600 compounds (Berdy, 2005). As a result, *streptomycetes* have become the primary antibiotic-producing actinomycetes exploited by the pharmaceutical industry.

Streptomyces coelicolor A 3(2), is the best-known strain of secondary metabolite producer of streptomycetes. According to Zhu *et al.,* (2014), many secondary metabolites have been discovered from this strain such as actinohodin (ACT), undecylprodigiosin (RED), calcium-dependent antibiotic (Cda), and the plasmid-encoded methylenomycin (Mmy). Besides, *S. coelicolor* genome sequence still revealed many previously unidentified biosynthetic gene clusters including one for a likely antibiotic called cryptic polyketide (Cpk) even after 50 years of research on it. A sequence study on antibiotic gene clusters and the complete genome of *S. coelicolor* revealed that such microorganisms are probably capable of producing a greater number of secondary metabolites (Higginbotham & Murphy, 2010).

ACTINORHODIN (ACT) AND UNDECYLPRODIGIOSIN (RED)

S. coelicolor synthesizes two chemically distinct pigments which are generally regarded as secondary metabolites which are actinorhodin (ACT), a diffusible red-blue pH indicator and undecylprodigiosin (RED), a red cell-wall associated compound (Rudd & Hopwood, 1980). During the past thirty years, researchers have shown an increased interest in RED compounds due to their immunosuppressive and anticancer properties in addition to antimicrobial activities. Meanwhile, ACT compound exhibit antibacterial activity against gram-positive cells (Mak, Xu & Nodwell, 2014)

Actinorhodin is an aromatic polyketide synthesized by enzymes encoded in a 22-kb gene cluster. The gene cluster responsible for actinorhodin production contains the biosynthetic enzymes and genes responsible for export of the antibiotic. The actinorhodin biosynthetic cluster also encodes a pathway-specific activator (actII-orf4) that activates the biosynthetic genes. This activator gene is in turn subject to the action of global regulators that can either activate or repress its expression (Craney, Ahmed & Nodwell, 2013). Furthermore, their production occurs using a type II polyketide synthase (PKS). The formation of actinorhodin started as the carbon backbone is produced entirely from fatty acids precursors, acetyl-CoA and malonyl-CoA in primary metabolism.

Meanwhile, undecylprodigiosin is a red pigmented, cell wall-associated antibiotic that belongs to a group of polypyrrole bioactive compounds called prodiginines (Luti & Yonis, 2014) which is directed by a 30-kb gene cluster. Two pathway-specific transcriptional activators involved for the activation of the prodiginine gene are RedZ and RedD. In the pathway, RedZ functions as a direct activator of RedD which then acts on the biosynthetic genes (Craney, Ahmed & Nodwell, 2013).

A study has been conducted with aimed to determine the relationship between secondary metabolite production and growth media composition. As a result, it shows that Act produced mainly in the stationary phase of batch cultures grown with glucose and sodium nitrate as the sources of carbon and nitrogen. Meanwhile, Red accumulated during the exponential phase. The production of both pigments were sensitive to the levels of ammonium and phosphate in the medium (Hobbs *et al.,* 1990).

Besides, several studies have been done on deletion of the coding region of the ppGpp synthetase gene, relA in *Streptomyces celicolor* A3 (2) correspond with antibiotic production. They noted that there is correlation between ppGpp synthetase gene, relA and the onset of undecylprodigiosin (Red) and actinorhodin (Act) production, leading to the suggestion that ppGpp plays a central role in triggering antibiotic synthesis (Chakraburtty *et al.,* 1996).

Studies of batch cultures, some of which were subjected to amino acid starvation, indicated a correlation between ppGpp synthesis and transcription between pathway specific regulatory genes for Red and Act (the two pigmented antibiotics made by the strain). The relA null mutant was grown at the same rate as the parental strains resulting in depletion production of both Act and Red under condition of nitrogen limitation, but appeared to produce normally under other conditions (Chakraburtty, R., & Bibb, M. 1997). This indicates that actinorhodin and undecylprodigiosin cannot be produced due to ppGpp synthetase gene, relA cannot work at its best under amino acid starvation.

GLUCOSE-6-PHOSPHATE DEHYDROGENASE (G6PDH) ASSAY

Previously, many researches had proved that the production of secondary metabolites depends on precursors supplement from primary metabolism. For instance, in 2012, a study was conducted by Wentzel *et al,* to find the relationship between carbon fluxes towards biomass formation and antibiotic production by changing carbon and nitrogen sources or varying initial seeding volumes of cells in cultivation media

(Cheng *et al.*, 2013). Both studies had revealed that the reaction related to the amino acid pathway helped in concentrating fluxes towards the biosynthesis of various precursors required for synthesizing secondary metabolites.

Following this, the recent study has been conducted by targeting pentose phosphate pathways to improve the production of secondary metabolites (Actinorhodin and Undecylprodigiosin). As mentioned by Fan *et al.,* (2016), pentose phosphate pathway plays an important role in secondary metabolite production and is considered as the sources of precursors.

$$G6PDH + G6P + NAD \rightarrow \quad \text{6-phospho-D-glucono-1,5-lactone} + NADPH$$

This is performed by maximizing the conversion of the first enzyme of the pathway, glucose-6-phosphate dehydrogenase (G6PDH) by supplying an adequate number of substrates which are glucose-6-phosphate (G6P) and nicotinamide adenine dinucleotide (NAD) to improve production of NADPH. As per suggested by Gunarson, Eliasson & Nielsen (2004), NADPH plays an important role in enhancement of secondary metabolites. NADPH is the reducing agent used in the process of making secondary metabolites, and the pentose phosphate pathway is one of the most important NADPH-producing pathways. The first enzyme of the pathway, glucose-6-phosphate dehydrogenase (G6PDH) is generally considered as an exclusive NADPH producer.

MATERIALS AND METHODS
BACTERIAL STRAINS
Streptomyces sp. K2-11 were taken from laboratory collections (Research Lab 3, Kulliyyah Science, IIUM Kuantan) which were isolated from mangrove sediment of Tanjung, Lumpur, Kuantan, Pahang.
.

PREPARATION OF MEDIA
Nitrogen limiting SMMS medium
As much as 2 g of Difco casamino acids, TES buffer ($5.68Gl^{-1}$) and Bacto agar were dissolved in distilled water. Then the pH was adjusted to 7.2 using 10 M NaOH prior autoclaving. The media with the following ingredients were added with specific amount: $NaH_2PO_4 + K_2H_2PO_4$ (50 Mm each, 10 mL per litre of culture), $MgSO_4.7H_2O$ (1 M, 5 mL per litre of culture), glucose (50% w.v, 18 mL per litre of culture). The trace elements that contain o.1 gL^{-1} each of $ZnSO_4.7H_2O$, $FeSO_4.7H_2O$, $MnCl_2.4H_2O$, $CaCl_2.6H_2O$ and NaCl. The solution was stored at $4^{0}C$ in a refrigerator.

CULTURING of *Actinomycetes*
All of the bacterial strains were grown on nitrogen limiting SMMS medium. The samples were incubated at $28^{0}C$, agitated at 120 rpm for fourteen days.

GLUCOSE-6-PHOSPHATE DEHYDROGENASE ASSAY
Preparation of Extracts
The method was performed according to the protocol by Borodina *et al.,* (2008). Cells used for activity assays were harvested after 67 h of growth in 200 ml of defined medium in a 1-liter flask equipped with a stainless-steel spiral. Cells were harvested by centrifugation and resuspended in buffer containing 50 mM TES, pH 7.2, 5 mM $MgCl_2$, 5 mM 2-mercaptoethanol, 50 mM $(NH_4)_2SO_4$, and 0.1 mM phenylmethylsulfonyl fluoride (buffer A). Lysozyme (add in the concentration) was used to break the cells.

G6PDH activity Assay

Glucose-6-phosphate dehydrogenase (G6PDH, EC 1.1.1.49) assays are based on the production of NADPH and were performed according to the protocol of Lessie and Wyk, (1972) and modified by Butler *et al.,* (2002). Both the consumption of NADH and the production of NADPH were measured spectrophotometrically at 340 nm. The crude lysates were applied to the G6PDH activity assay using supplied substrates (G6P and NAD). The assay was performed in a 96-well plate for two minutes enabling simultaneous analysis of a large number of samples.

$$G6PDH + G6P + NADP \rightarrow \quad 6\text{-phospho-D-glucono-1,5-lactone} + NADPH$$

RESULTS AND DISCUSSION
EXTRACTS PREPARATION
Five genera of Actinomycetes which are *Streptomyces, Micromonospora, Nocardia, Nocardiopsis* and *Rodhococcus* were taken from laboratory collections. These microbes have been identified and known to produce antimicrobial activity. All of the isolates were grown on nitrogen limiting SMMS medium. However, due to time constraint, only *Streptomycetes* was chosen to be assayed for secondary metabolite production. The *Streptomycetes* was grown on SMMS plate for five days and was subcultured into SMMS broth for another three days according to the protocol of Borodina *et al.,* (2008). Then, the cells were harvested by using centrifugation and resuspended in a buffer and then repeated for three times. This is to make sure the 90 % of cells was lysed and released the protein. Phenylmethylsulfonyl fluoride which is known as a serine protease inhibitor was included in the buffer to prevent the protein degradation.

GLUCOSE-6-PHOSPHATE DEHYDROGENASE ASSAYS
The crude lysates were applied to the G6PDH activity assay using supplied substrates (G6P and NADP). The assay was performed in a 96-well plate that enabling simultaneous analysis of a large number of samples. The reaction was monitored by measuring the absorbance at 340 nm for two minutes and reduced cofactor, NADPH were absorbed readily at this wavelength.

The reaction rates measured at different substrates and protein concentration were shown in Figure 4.1. In order to obtain the best activity curve for the given condition, seven samples of different crude lysates concentrations were prepared (100 μL, 50 μL, 25 μL, 12.5 μL, 6.25 μL, 3.125 μL, and 1.5625 μL). Then, all the samples were subjected to various substrate concentrations to screen for the best enzyme activity. In this study, eight substrate concentrations were chosen to be tested with different enzyme concentrations (2 μM, 5 μM, 10 μM, 20 μM, 30 μM, 40 μM, 50 μM and 60 μM). The results show that the rate of reaction of various substrate concentrations were increased as the enzyme concentration increased. Reaction with 20 μM of substrate has the highest enzyme activity. Meanwhile, the least enzyme activity was shown in reaction with 50 μM of substrate for all enzyme concentration tested.

Figure 4.1 shows that at higher concentrations of crude lysates specifically 100 μM, 50 μM and 25 μM, the reaction was not stable when subjected to lower concentration of substrates (2 μM, 5 μM, 10 μM, 20 μM). However, the reactions started to increase at substrate concentration 30 μM to 60 μM. These conditions were contradicted with the reaction shown by lower concentrations of crude lysates (12.5 μM, 6.25 μM, 3.125 μM and 1.5625 μM) where the reaction increased slightly at lower concentration of substrates and decreased with in the presence of high concentration of substrate. Hence, it can be seen that higher enzyme and substrate concentration will increase the activity whereas lower concentration of enzyme with higher concentration of substrate will reduce the activity.

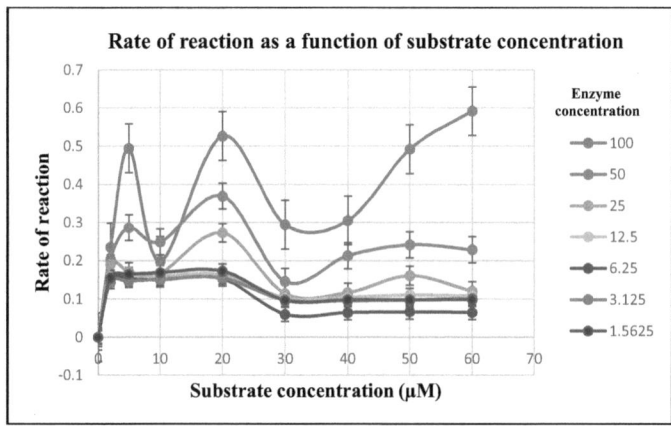

Fig. 4.1: Measurement of enzyme activities from crude lysates produced at wavelength 340 nm with different substrate concentrations. All readings have been normalized with control

Overall, it can be concluded that the enzyme activity works at its best with increasing enzyme's concentrations as well as substrate. However, a better assay could be conducted by using a purified enzyme. According to Sharma and Chand, (2012), purified protein exhibits better activity readings compared to crude enzymes. This might be due to protein impurities present in the reaction which may interfere with the absorbance readings.

According to Bisswanger (2014), there are several factors that may affect the assay other than pH, temperature and ionic strength. For example, the actual concentrations of all assay components. This may contribute to the deviations from the optimum conditions of the protein which causes a reduction of the activity. For instance, enzyme reactions dependent on ATP need Mg^{2+} as essential counter ions. The assay mixture will become limiting if only ATP without Mg^{2+} were added even in sufficient concentration especially if complexing compounds like inorganic phosphates or EDTA are present. In this study, this also could be considered as a contributing factor on the fluctuated readings. This physicochemical property of G6PDH enzymes needs further study for a better assay condition.

CONCLUSION
This preliminary attempt to optimize glucose-6-phosphate dehydrogenase activity assay was encouraging. Even though glucose-6-phosphate dehydrogenase activity assay was not fully optimized, there is some knowledge that we can still perceive out of this project. One of the knowledge was this enzyme is an allosteric which does not obey the Michealis –Menten kinetics due to the presence of multiple binding sites. It is believed, with improvement of certain factors like using purer enzymes, the study could offer more promising results. In addition, this protein has higher potential towards secondary metabolite production through the formation of NADPH as G6PDH is generally considered as NADPH producer through pentose phosphate pathway (PPP). Nevertheless, an intense research on the physical and physicochemical properties of G6PDH should be conducted for a better understanding of the whole enzymatic reaction.

REFERENCES

Barka, E. A., Vatsa, P., Sanchez, L., Gaveau-Vaillant, N., Jacquard, C., Klenk, H. P., ... & van Wezel, G. P. (2016). Taxonomy, physiology, and natural products of Actinobacteria. *Microbiology and Molecular Biology Reviews*, *80*(1), 1-43.

Berdy, J. (2005). Bioactive microbial metabolites. *Journal of Antibiotics*,*58*(1), 1.

Bisswanger, H. (2014). Enzyme assays. *Perspectives in Science*, *1*(1), 41-55.

Borodina, I., Siebring, J., Zhang, J., Smith, C. P., van Keulen, G., Dijkhuizen, L., & Nielsen, J. (2008). Antibiotic overproduction in Streptomyces coelicolor A3 (2) mediated by phosphofructokinase deletion. *Journal of Biological Chemistry*, *283*(37), 25186-25199.

Brockman, I. M., Prather, K. L. J., & Gupta, A. (2017). Dynamic Knockdown of Central Metabolism for Redirecting Glucose-6-Phosphate Fluxes. *U.S. Patent No. 20,170,130,210*. Washington, DC: U.S. Patent and Trademark Office.

Butler, M. J., Bruheim, P., Jovetic, S., Marinelli, F., Postma, P. W., & Bibb, M. J. (2002). Engineering of primary carbon metabolism for improved antibiotic production in Streptomyces lividans. *Applied and environmental microbiology*, *68*(10), 4731-4739.

Craney, A., Ahmed, S., & Nodwell, J. (2013). Towards a new science of secondary metabolism. *The Journal of antibiotics*, *66*(7), 387-400.

Chaudhary, H. S., Soni, B., Shrivastava, A. R., & Shrivastava, S. (2013). Diversity and Versatility of Actinomycetes and its Role in Antibiotic Production. *Journal of Applied Pharmaceutical Science, 3*(8), 83-94.

Chakraburtty, R., White, J., Takano, E., & Bibb, M. (1996). Cloning, characterization and disruption of a (p)ppGpp synthetase gene (relA) of Streptomyces coelicolor A3 (2). *Molecular microbiology*, *19*(2), 357-368.

Chakraburtty, R., & Bibb, M. (1997). The ppGpp synthetase gene (relA) of Streptomyces coelicolor A3 (2) plays a conditional role in antibiotic production and morphological differentiation. *Journal of Bacteriology*, *179*(18), 5854-5861.

Cheng, J. S., Liang, Y. Q., Ding, M. Z., Cui, S. F., Lv, X. M., & Yuan, Y. J. (2013). Metabolic analysis reveals the amino acid responses of Streptomyces lydicus to pitching ratios during improving streptolydigin production. *Applied microbiology and biotechnology*, *97*(13), 5943-5954.

Das, S., Lyla, P. S., & Khan, S. A. (2008). Distribution and generic composition of culturable marine actinomycetes from the sediments of Indian continental slope of Bay of Bengal. *Chinese Journal of Oceanology and Limnology*, *26*(2), 166-177.

Doelle, H. W. (2014). Aerobic Respiration. *Bacterial metabolism* (pp. 364). Academic Press.

Dyson, P. (2011). *Streptomyces: molecular biology and biotechnology*. Horizon Scientific Press.

Ethiraj, T., Revathi, R., Thenmozhi, P., Saravanan, V. S., & Ganesan, V. (2011). High performance liquid chromatographic method development for simultaneous analysis of doxofylline and montelukast sodium in a combined form. *Pharmaceutical methods*, *2*(4), 223-228.

Fan, Y., Hu, F., Wei, L., Bai, L., & Hua, Q. (2016). Effects of modulation of pentose-phosphate pathway on biosynthesis of ansamitocins in Actinosynnema pretiosum. *Journal of biotechnology*, *230*, 3-10.

Goodfellow, M., & Williams, S. T. (1983). Ecology of actinomycetes. *Annual Reviews in Microbiology*, *37*(1), 189-216.

Gunarson, N., Eliasson, A., & Nielsen, J. (2004). Control of fluxes towards antiobiotics and the role of primary metabolism in production of antiobiotics. *Advance Biochemica. Engineering Biotechnology.*, *88*, 137-178.

Higginbotham, S. J., & Murphy, C. D. (2010). Identification and characterisation of a Streptomyces sp. isolate exhibiting activity against methicillin-resistant Staphylococcus aureus. *Microbiological Research*,*165*(1), 82-86.

Hobbs, G., Frazer, C. M., Gardner, D. C., Flett, F., & Oliver, S. G. (1990). Pigmented antibiotic production by Streptomyces coelicolor A3 (2): kinetics and the influence of nutrients. *Journal of General Microbiology*, *136*(11), 2291-2296.

Kämpfer, P. (2015). Streptomyces. *Bergey's Manual of Systematics of Archaea and Bacteria*, 1-414.

Kun, L. Y. (2003). Screening for antimicrobial products. *Microbial biotechnology:principles and applications*. (pp. 13). World Scientific.

Lessie, T. G., & Vander Wyk, J. C. (1972). Multiple forms of Pseudomonas multivorans glucose-6-phosphate and 6-phosphogluconate dehydrogenases: differences in size, pyridine nucleotide specificity, and susceptibility to inhibition by adenosine 5′-triphosphate. *Journal of bacteriology*, *110*(3), 1107-1117.

Lo, C. W., Lai, N. S., Cheah, H. Y., Wong, N. K. I., & Ho, C. C. (2002). Actinomycetes isolated from soil samples from the Crocker Range Sabah. *ASEAN Review on Biodiversity and Environmental Conservation*.

Luti, K. J. K., & Yonis, R. W. (2014). An induction of Undecylprodigiosin Production from Streptomyces coelicolor by Elicitation with Microbial Cells Using Solid State Fermentation. *Iraqi Journal of Science*, 55(4A), 1553-1562.

Madigan, M. T., Martinko, J. M., Dunlap, P. V., & Clark, D. P. (2008). Brock Biology of microorganisms 12th edition. *International Microbiology*, *11*, 65-73.

Mak, S., Xu, Y., & Nodwell, J. R. (2014). The expression of antibiotic resistance genes in antibiotic-producing bacteria. *Molecular microbiology*, *93*(3), 391-402.

Rudd, B. A., & Hopwood, D. A. (1980). A pigmented mycelial antibiotic in Streptomyces coelicolor: control by a chromosomal gene cluster.*Microbiology*, *119*(2), 333-340.

Sharma, P. K., & Chand, D. (2012). Purification and Characterization of Thermostable Cellulase Free Xylanase from Pseudomonas sp. XPB-6.

Solanki, R., Khanna, M., & Lal, R. (2008). Bioactive compounds from marine actinomycetes. *Indian journal of microbiology*, *48*(4), 410-431.

Ventura, M., Canchaya, C., Tauch, A., Chandra, G., Fitzgerald, G. F., Chater, K. F., & Sinderen, D. (2007). Genomics of Actinobacteria: tracing the evolutionary history of an ancient phylum. *Microbiology and Molecular Biology Reviews*, *71*(3), 495-548.

Weber, T., Charusanti, P., Musiol-Kroll, E. M., Jiang, X., Tong, Y., Kim, H. U., & Lee, S. Y. (2015). Metabolic engineering of antibiotic factories: new tools for antibiotic production in actinomycetes. *Trends in biotechnology*, *33*(1), 15-26.

Wentzel, A., Bruheim, P., Øverby, A., Jakobsen, Ø. M., Sletta, H., Omara, W. A. & Ellingsen, T. E. (2012). Optimized submerged batch fermentation strategy for systems scale studies of metabolic switching in Streptomyces coelicolor A3 (2). *BMC systems biology*, *6*(1), 59.

Zhu, H., Sandiford, S. K., & van Wezel, G. P. (2014). Triggers and cues that activate antibiotic production by actinomycetes. *Journal of industrial microbiology & biotechnology*, *41*(2),371-386.

Cultivation vs the 'omics' approach for microbial bioprospecting in the 21st century: Coastal environment in Malaysia

Suhaila Mohd Omar [1]*

[1]*Dept. of Biotechnology, Kulliyyah of Science, International Islamic University Malaysia*
Corresponding author: osuhaila@iium.edu.my

ABSTRACT

The coastal environment is the habitat of diverse functionally important marine microorganisms. Among the valuable characteristics of the microorganisms for bioprospecting studies are not limited to tolerant toward rapid and repeated fluctuations in temperature, sunlight, salinity, wave action, ultraviolet radiation, and periods of drought. On the other hand, microorganisms that are living the epiphytic, epibiotic, and symbiotic lifestyles produce specific toxins, signalling molecules, and other secondary metabolites due to their defence and signalling mechanism. The traditional and innovative cultivation method is still relevant in bioprospecting studies while the 'omics' approaches offer an extensive gateway towards microorganism diversity and function. Therefore, this minireview focus on the challenges, strategies and the success of microbial bioprospecting studies in the context of Malaysia coastal environment via cultivation and 'omics' approach.

Keywords: Omics; microbs; symbiont; microbe culture

INTRODUCTION

The total 4,800 km coastline of Malaysia comprises two distinctly different physical formations including mangrove-fringed mudflats and sandy beaches that host distinct, unique and spectacular biodiversity ((MYBIS, 2015). The straight sandy formations are prevalent in the northeast coast of Peninsular Malaysia while the south comprises a series of hook- or spiral-shaped bays. Meanwhile, the west coast of Peninsular has limited areas of pocket sandy beaches and mostly is made up of muddy formations. The coastline in Sarawak and Sabah comprises almost equally divided sandy beaches and mud coast (Abdullah, 1993). The earliest report on marine diversity dated back in 1849 comprises fish diversity catalogue (Cantor, 1849). In comparison to fishes, reptiles, mammals, invertebrates, sea cucumber (Holothuroid) and seagrasses, the detailed accounts on other marine organisms especially microorganisms are still lacking (Mazlan et al., 2005). Moreover, the known Coral Triangle, which includes the reefs of Indonesia, the Philippines and Malaysia, made up 76% of all known coral species and hosts 37% of all known coral reef fish species worldwide (Burke, 2011). The exceptional biodiversity of the marine habitats offers a valuable opportunity for bioprospecting. This minireview highlights on the Malaysia coastal marine microbial biodiversity and bioprospecting studies via cultivation and 'omics' approach.

Coastal environment as habitat of functionally important marine microorganisms

Bioprospecting is a targeted and systematic exploration for components, bioactive compounds or genes within living organisms. This may include all kinds of organisms; microorganisms like bacteria, fungi and viruses and larger organisms such as sea plants, shellfish and fish (Ministry of Fisheries and Coastal Affairs, 2009; Mossop, 2015). The marine environment covers more than 70% of the Earth's surface and contains 97.5% of the water of our planet. Microorganisms represent the majority of the rich and diverse life of the marine habitat. Among the environmental factors that distinguish marine microbial community composition compared to the terrestrial environment is the salinity (Vogel et al., 2020). The complex coastal microbial communities also play important roles in the regulation of the biogeochemical cycling at the land-sea interface thus include all domains of life and form a network that connects the water column and sediment (Fuhrman et al., 2015; Moulton et al., 2016). Microorganisms from intertidal zones must be able to strive in the extreme condition such as rapid and repeated fluctuations in temperature, sunlight, salinity, wave action, ultraviolet radiation, and periods of drought (McKew et al., 2011).

From a biotechnological perspective, the group of microorganisms those living under epiphytic, epibiotic, and symbiotic lifestyles are also incomparable pool reservoir due to their specific competition and defence strategies characteristic of surface-associated microorganisms, such as the production of toxins, signaling molecules, and other secondary metabolites (Gonzalez et al., 2016). Sponges and corals are examples of habitats where symbiotic associations of microorganisms can be found in sponges and corals as well as with marine invertebrates (Amelia et al., 2020; Hanani et al., 2015). The end product from the bioprospecting activities could be a purified molecule that is produced biologically or synthetically or the entire organism. Even though marine bioprospecting is not an industry in the traditional sense, the potential to acquire new compounds for use in many different industries is the interesting driving force. Over the years, novel and more complex approaches have been developed and utilized to study marine microbial biodiversity and their biotechnological potential.

Methods to explore marine microbial biodiversity and potential application: Cultivation approach
The low cultivability of marine microbes is well known and referred to as the 'great plate count anomaly' (Staley & Konopka, 1985) due to the difference between the number of colonies that developed on laboratory medium and the total number of bacteria that could be counted by epifluorescence microscopy of DAPI-stained samples. The metabolic potential of microbes in the laboratory or ecosystem function can only be corroborated through studies of cultivated organisms (Prakash et al., 2013). Therefore, the isolation, characterization, and preservation of novel microbes are a requisite for the future growth of bioprospecting from the marine environment. Table 1 illustrates the list of some of the cultivated microbes from different Malaysia coastal environments for the past 20 years and their potential application. *Alphaproteobacteria* and *Gammaproteobacteria* dominated the culture collection. Some of the researchers use half strength of the common marine agar composition as the effort to increase the isolation of novel strain (Kuek et al., 2016). Diversity of the medium formula used for cultivation (Law et al., 2019) as well as wet and dry heat pretreatment also increase recovery of novel Actinomycetes (Abdul Malek et al., 2015). Bacteria belonging to the genus *Streptomyces* have been acknowledged as the producers of many bioactive compounds, which makes them to be important microorganisms for secondary metabolites with potential anticancer, antimicrobial roles because of their cytotoxic properties (Law et al., 2019). The potential application of the isolates ranges from enzyme discovery (Cheng et al., 2020; Dinesh et al., 2017; Naresh et al., 2019; Omar et al., 2017; Yasim, 2018), bioremediation (Hanani et al., 2015; Kuek et al., 2016), antibacterial and antifungal (Zainal Abidin et al., 2016). The prevalence of antibiotic resistance bacteria and its high impact on human health urge the need of searching for new natural products that could therefore remedy this problem, especially from the marine environment (Jalal et al., 2012). Most of the isolates were recovered via modification of standard plating technique which can recover a very small proportion, 0.001–1% of the total assemblage (Staley & Konopka, 1985). Cultivation followed by high-throughput screening for specific functions is another strategy for researchers with the advanced facilities to increase positive hits (Law et al., 2019).

Table 1: Selected Microbial Bioprospecting via Cultivation Approach in the Coastal Environment, Malaysia (2000-2020)

No.	Sampling Location	Microbial Strains	Potential Application	Ref
1	Marine resources (horseshoe crab from Sabah, jellyfish from Sarawak, mollusc and marine sediment from	*Bacillus, Chryseomicrobium, Photobacterium, Pseudoalteromonas, Ruegeria, Shewanella,*	Enzyme: amylase, lipase and protease	(Cheng et al., 2020)

	Kelantan and marine water from Terengganu) .	*Solibacillus, Tenacibaculum and Vibrio.*		
2	Mangrove forest soil, Tanjung Lumpur, Pahang	*Verrucosispora* sp. K2-04	Enzyme: xylanase	(Omar et al., 2017)
3	Estuarine mangrove sediment of Matang Mangrove Forest	*Mangrovimonas xylaniphaga sp.* nov.	Enzyme: Xylanase	(Dinesh et al., 2017)
4	Mangrove roots collected in Tanjung Piai, Johor	*Exiguobacterium* sp. CN10	Enzyme for lignocellulosic biomass degradation	(Yasim, 2018)
5	Mangrove soil from northern states of Malaysia (Perlis, Kedah, Pulau Pinang and Perak).	*Bacillus subtilis* KB01; *Anoxybacillus* sp. UniMAP-KB02, KB03, KB04 KB05, KB06; *Paenibacillus dendritiformis* UniMAP-KB01	Thermophilic cellulase	(Naresh et al., 2019)
6	South China Sea and along the coastline of Peninsular Malaysia and Borneo	*Alphaproteobacteria: Caulobacteraceae, Phyllobacteriaceae, Rhodobacteraceae and Rhodospirillaceae,* *Betaproteobacteria: Alcaligenes sp.* *Gammaproteobacteria: Aeromonadaceae, Pseudoalteromonadaceae, Shewanellaceae, Pseudomonadaceae and Vibronaceae*	Bioremediation, nitrogen fixing and sulphate reduction	(Kuek et al., 2016)
7	Pulau Kapas Beach and Pantai Batu Burok, Terengganu.	NA	Antibacterial activities	(Mazalan et al., 2012)
8	Mangrove soil in Kuching, Sarawak	*Streptomyces* sp.	Bioactive potentials-in relation to antioxidant and cytotoxic activities	(Law et al., 2019)
9	Mangrove forest soil, Tanjung Lumpur, Pahang	*Streptomyces mangrovisoli* sp. nov	Antioxidant identied as Pyrrolo [1,2-a]pyrazine-1,4-dione,hexahydro	(Ser et al., 2015)

10	Mangrove forest soil, Tanjung Lumpur, Pahang	*Streptomyces*-like and *Micromonospora*-like isolates	Antibacterial and antifungal	(Zainal Abidin et al., 2016)
11	Marine sponge (*Gelliodes* sp.) collected from coastal area of Kuantan	*Bacillus* sp.	Bioremediation-haloalkanoic acid (3-chloropropionic acid (3CP)-degrading activities	(Hanani et al., 2015)
12	Marine sediment of Songsong Island, Kedah, Malaysia.	18 *Streptomyces* isolates	Anti-infectives	(Fatin et al., 2017)

Omics and meta omics approach
The innovative breakthroughs in genome sequencing, bioinformatics, and analytic tools such as liquid and gas chromatography and mass spectrometry, along with high-throughput technologies has promoted the advances in "omics" technologies (genomics, transcriptomics, proteomics, and metabolomics). In comparison to genomics that study specific isolate, metagenomics is a technique that involves sequencing DNA from the genomes of all organisms present in a particular sample and has become a common method for the study of microbiome population structure and function. Through this approach, the genes and pathways from the whole microbiome can be determined. The metagenomic methods can be classified based on sequencing of metagenomes and bioinformatic analysis or functional expression of metagenomic libraries to identify genes or gene clusters of interest. Since there is no need to isolate or cultivate the microorganisms, directly extracted DNA provides information on the metabolic and functional capacity of a specific cultivable and uncultivable microbial community (Simon & Daniel, 2011). Metagenomics goes hand in hand with next generation sequencing and high-performance supercomputing, thus enabling broad access to microorganism diversity and function (Knight et al., 2012). On the other hand, metatranscriptomics helps to explain which metabolic pathways and genes are expressed in a given place at a given time. Both genomic DNA and total RNA libraries can be prepared and sequence in parallel following a proper sample handling and nucleic acids extraction protocol (Mason et al., 2012). Another two approaches, metaproteomics is the quantification of protein or peptide levels, while metabolomics is related to the investigation of small-molecule metabolites. Among the four, genomics and metagenomics are the most popular methods used to study coastal microbiome in Malaysia. As of the time of writing, there is no report on metaproteomics or metabolomics-based study found.

Genomics approach
Table 2 showed the examples of the successful application of genomics towards several bacterial isolates for determination of enzyme and secondary metabolite gene clusters. Genomic study of *Catenovulum*-like strain CCB-QB4 and *Aureispira* sp. CCB-QB1 from Penang coastal environment highlighted arachidonic acid biosynthesis (Lau et al., 2019a) and polyunsaturated fatty acid and diterpenoid biosynthesis pathways (Furusawa et al., 2015) respectively. Another two strains from Hulu Selangor, *Vibrio variabilis* strain T01 (Mohamad et al., 2016) and *Vibrio sinaloensis* 147 (Mohamad et al., 2017) reveal the quorum sensing properties. Meanwhile, *Streptomyces* sp. MUSC 125 and *Yangia* sp. strain CCB-MM3 from mangrove environment were confirmed with pathway and genes related to antioxidant (Ser et al., 2016) and polyhydroxyalkanoate copolymer production (Lau et al., 2017) respectively. Data mining of the genomic sequences for the six bacteria belonging to the genus *Novosphingobium* from the National Center for Bioinformatic Information (NCBI) database also provides useful insights towards genes related to marine adaptation, cell-cell signaling and bioremediation (Gan et al., 2013).

Table 2: Selected Microbial Bioprospecting via Genomic Approach in the Coastal Environment, Malaysia (2000-2020)

No	Sampling location	Microbial strain	Potential application	Ref.
1.	Coastal area of Penang	*Catenovulum*-like strain CCB-QB4	Agarase	(Lau et al., 2019b)
2.	Coastal area of Penang	*Aureispira* sp. CCB-QB1	Linoleoyl-CoA desaturase, the key gene in arachidonic acid biosynthesis.	(Furusawa et al., 2015)
3.	Coastal waters in Hulu Selangor	*Vibrio variabilis* strain T01	Quorum sensing	(Mohamad et al., 2016)
4.	Morib Beach, Hulu Selangor.	*Vibrio sinaloensis* T47	Quorum sensing	(Mohamad et al., 2017)
5.	Mangrove soil in the east coast of Peninsular Malaysia	*Streptomyces* sp. MUSC 125	Antioxidant properties	(Ser et al., 2016)
6.	Soil sediment in the estuarine Matang Mangrove Forest Reserve	*Yangia* sp. strain CCB-MM3	Pathway for production of propionyl-CoA and gene cluster for PHA production	(Lau et al., 2017)
7.	NCBI database	six bacteria belonging to the genus *Novosphingobium*	Marine adaptation, cell-cell signalling and bioremediation	(Gan et al., 2013)

Metagenomic Approach

The ability to profile diverse microbial communities using next-generation sequencing (NGS) boosted the interest in microbiome research. Through this culture-free and high-throughput technology, the identification and comparison of whole microbial communities, also known as metagenomics can be accomplished. Metagenomics typically encompasses two particular sequencing strategies: amplicon sequencing, most often of the 16S rRNA gene as a phylogenetic marker; or shotgun sequencing, which captures the complete breadth of DNA within a sample (Morgan & Huttenhower, 2012).

There is a limited report of 'omics' approach study in Malaysia coastal microbiome. As shown in Table 3, most of the studies were restricted to the bioinformatic analysis of the 16S rRNA amplicon sequencing and shotgun metagenomic sequencing data. Both of the sequencing strategies have their advantage and application. The use of the 16S ribosomal RNA gene as a phylogenetic marker has proven to be an efficient and cost-effective strategy for microbiome analysis and even allows for the prediction of functional content based on taxon abundances. Alternatively, scientist may go for a direct experimental approach to unravel novel biochemical function of unknown protein by screening for purified proteins or metagenomic gene libraries that use either *E. coli* (Lee et al., 2015) or lambda phage as cloning host (Popovic et. al., 2017). For example, the abundance of sulfur-degrading bacteria in a benthic bacterial community of shallow sea sediment off-Terengganu coast of the South China Sea was detected through this strategy. The physical-geochemical analysis revealed that the surveyed areas contained sulphur, oil, grease, gasoline, diesel, and mineral oil, which suggests the effect of environment condition towards the prevalence of sulphur-degrading bacteria community growth in the northeast area of the surveyed area (Marziah et al., 2016). However, there is an issue on the vulnerability of this protocol to biases through sample preparation and sequencing errors. Furthermore, 16S rRNA gene amplicon sequencing is typically limited to taxonomic classification at the genus level depending on the database and classifiers used and provides only limited functional information (Morgan & Huttenhower, 2012). On the other hand, shotgun metagenomics offer both phylogenetic surveys as well functional gene composition of microbial communities (Thomas et al., 2012). In the Matang Mangrove Forest metagenome of the Productive Zone, the microbial community was overabundant in genes related to carbohydrate metabolism, especially enzymes involved in the degradation and utilization of polysaccharides from plant cell walls. Functional analysis focusing on carbohydrate degrading enzymes revealed an array of enzymes involved in hemicellulose, cellulose and pectin utilization enzymes (Priya et al., 2018). The downside of shotgun metagenomics that limited its wider usage is its relatively high costs and more demanding bioinformatic requirements (Morgan & Huttenhower, 2012; Rausch et al., 2019).

Apart from relying on previous sequence knowledge for identification, sequence-based metagenomics allows the identification of a huge number of genes encoding putative functions without guarantee that the genes will be successfully expressed in the heterologous host. On the other hand, even though the functional screening of metagenomics libraries may offer novel findings, the relatively high cost of imported molecular kits and cloning vectors, laborious and potentially low hits in the screening process (Kennedy et al., 2008), could be the reason that this approach is not attractive enough for the local researchers.

Table 3: Selected Metagenomics Study in Malaysia (2000-2020)

No	Sampling Location	Sequencing Approach/Platform	Ref.
1.	Along the coast of Borneo, Malaysia, and the Philippines	Shotgun metagenomic sequencing/Illumina HiSeq2000	(Song et al., 2017)
2.	Surface Seawater of Georgetown Coast	Shotgun sequencing/ (Miseq) Ilumina	(Arumugam et al., 2013)
3.	Seawater at the surface of the littoral zone was collected from an estuary in Sabak Bernam, and a fishing village in Sekinchan, Selangor	16s rNA gene amplicon sequencing	(Chan & Chong, 2014)
4.	Sediment off-Terengganu coast of the South China Sea	16s rRNA amplicon sequencing (Illumina) Miseq	(Marziah et al., 2016)
5.	Soil of the Virgin Jungle Forest and the harvested Productive Zone of Matang Mangrove Forest Reserve	Shotgun metagenomics/ Ilumina HiSeq2500	(Priya et al., 2018)
6.	Seawater of South China Sea continuum (Rajang river and estuaries lead to the sea)	16s rRNA amplicon sequencing/ Illumina	(Sien Aun Sia et al., 2019)

7.	Sponges (*Aaptos aaptos* and *Xestospongia muta*) from Bidong and Redang islands.	16S rRNA amplicon sequencing/ Illumina HiSeq2500	(Amelia et al., 2020)

CONCLUSION

It is important to emphasize that a 16S rRNA gene sequence alone is probably not sufficient to uniquely identify any microbe in the environment. However, the data can be utilised to develop a targeted and improved cultivation media and technique. Furthermore, the development of more versatile vectors, host strain engineering and high throughput, cheap functional screening assays could improve the low hit rate associated with functional metagenomics. The combination of cultivation, sequence and functional-based approach, followed by biochemical and pharmaceutical studies will potentially unravel various components, bioactive compounds or genes from the enormous majority of the uncultivated microorganisms in the environment.

REFERENCES

Abdul Malek, N., Zainuddin, Zarina, Chowdhury, A.J.K, Zainal Abidin, Z (2015). Diversity and antimicrobial activity of mangrove soil actinomycetes isolated from Tanjung Lumpur, Kuantan. *Jurnal Teknologi, 77* (25). , 0 pp. 37-43. ISSN 0127–9696

Abdullah, S. (1993). *Coastal Developments in Malaysia-Scope, Issues and Challenges.* https://www.water.gov.my/jps/resources/auto%20download%20images/5844e2da4907f.pdf

Amelia, T. S. M., Lau, N.-S., Amirul, A.-A. A., & Bhubalan, K. (2020). Metagenomic data on bacterial diversity profiling of high-microbial-abundance tropical marine sponges *Aaptos aaptos* and *Xestospongia muta* from waters off Terengganu, South China Sea. *Data in Brief, 31*, 105971. https://doi.org/10.1016/j.dib.2020.105971

Arumugam, R., Chan, X.-Y., & Woh Choo, S. (2013). Metagenomic analysis of Microbial Diversity of Tropical Sea Water of Georgetown Coast, Malaysia. https://www.researchgate.net/publication/287558965

Burke, L. (2011). *Reefs at risk revisited* (L. Burke, K. Reytar, M. Spalding, & A. Perry, Eds.). World Resources Institute.

Cantor, T. (1849). *Catalouge of Malayan Fishes.*

Chan, K.-G., & Chong, T.-M. (2014). Prevalence of Unclassified Bacteria in Tropical Coastal Waters of Malaysia Revealed by Metagenomic Approach. *Genome Announcements, 2*(3). https://doi.org/10.1128/genomeA.00419-14

Cheng, T. H., Ismail, N., Kamaruding, N., Saidin, J., & Danish-Daniel, M. (2020). Industrial enzymes-producing marine bacteria from marine resources. *Biotechnology Reports, 27*, e00482. https://doi.org/https://doi.org/10.1016/j.btre.2020.e00482

Dinesh, B., Furusawa, G., & Amirul, A. A. (2017). Mangrovimonas xylaniphaga sp. nov. isolated from estuarine mangrove sediment of Matang Mangrove Forest, Malaysia. *Archives of Microbiology, 199*(1), 63–67. https://doi.org/10.1007/s00203-016-1275-8

Fatin, S. N., Boon-Khai, T., Shu-Chien, A. C., Khairuddin, M., & Abdullah, A. A. A. (2017). A marine actinomycete rescues *Caenorhabditis elegans* from *Pseudomonas aeruginosa* infection through restitution of Lysozyme 7. *Frontiers in Microbiology, 8*(NOV). https://doi.org/10.3389/fmicb.2017.02267

Fuhrman, J. A., Cram, J. A., & Needham, D. M. (2015). Marine microbial community dynamics and their ecological interpretation. *Nature Reviews Microbiology, 13*(3), 133–146. https://doi.org/10.1038/nrmicro3417

Furusawa, G., Lau, N.-S., Shu-Chien, A. C., Jaya-Ram, A., & Amirul, A.-A. A. (2015). Identification of polyunsaturated fatty acid and diterpenoid biosynthesis pathways from draft genome of *Aureispira* sp. CCB-QB1. *Marine Genomics, 19*, 39–44. https://doi.org/https://doi.org/10.1016/j.margen.2014.10.006

Gan, H. M., Hudson, A. O., Rahman, A. Y. A., Chan, K. G., & Savka, M. A. (2013). Comparative genomic analysis of six bacteria belonging to the genus *Novosphingobium*: Insights into marine adaptation, cell-cell signaling and bioremediation. *BMC Genomics, 14*(1). https://doi.org/10.1186/1471-2164-14-431

Gonzalez NB, C., Toquica JS, R., Kleine L, L., & Castano D, M. (2016). Epiphytic Bacteria of Macroalgae of the Genus *Ulva* and their Potential in Producing Enzymes Having Biotechnological Interest. *Journal of Marine Biology & Oceanography, 5*(2). https://doi.org/10.4172/2324-8661.1000153

Hanani, N. S., Naim, A. M., Tengku Abdul Hamid, T. H., Huyop, F., & Abdul Hamid, A. A. (2015). Isolation and identification of 3- Chloropropionic acid degrading bacterium from marine sponge (Vol. 77). www.jurnalteknologi.utm.my

Jalal, K. C. A., Akbar, B. John., Kamaruzzaman, B. Y., & Kathiresan, K. (2012). *Emergence of Antibiotic Resistant Bacteria from Coastal Environment – A Review. in Antibiotic Resistant Bacteria-A Continuous Challenge in the New Millennium.* InTech.

Kennedy, J., Marchesi, J. R., & Dobson, A. D. (2008). Marine metagenomics: strategies for the discovery of novel enzymes with biotechnological applications from marine environments. *Microbial Cell Factories, 7*(1), 27. https://doi.org/10.1186/1475-2859-7-27

Knight, R., Jansson, J., Field, D., Fierer, N., Desai, N., Fuhrman, J. A., Hugenholtz, P., van der Lelie, D., Meyer, F., Stevens, R., Bailey, M. J., Gordon, J. I., Kowalchuk, G. A., & Gilbert, J. A. (2012). Unlocking the potential of metagenomics through replicated experimental design. *Nature Biotechnology, 30*(6), 513–520. https://doi.org/10.1038/nbt.2235

Kuek, F. W., Mujahid, A., Lim, P.-T., Leaw, C.-P., & Mueller, M. (2016). Diversity and DMS (P)-related genes in culturable bacterial communities in Malaysian coastal waters. *Sains Malaysiana, 45*(6), 915–931.

Lau, N.-S., Sam, K.-K., & Amirul, A. A.-A. (2017). Genome features of moderately halophilic polyhydroxyalkanoate-producing Yangia sp. CCB-MM3. *Standards in Genomic Sciences, 12*(1), 12. https://doi.org/10.1186/s40793-017-0232-8

Lau, N.-S., Tan, W. R., Furusawa, G., & Amirul, A.-A. A. (2019a). Complete genome sequence of the novel agarolytic Catenovulum-like strain CCB-QB4. *Marine Genomics, 43*, 50–53. https://doi.org/https://doi.org/10.1016/j.margen.2018.08.009

Lau, N.-S., Tan, W. R., Furusawa, G., & Amirul, A.-A. A. (2019b). Complete genome sequence of the novel agarolytic Catenovulum-like strain CCB-QB4. *Marine Genomics, 43*, 50–53. https://doi.org/https://doi.org/10.1016/j.margen.2018.08.009

Law, J. W. F., Chan, K. G., He, Y. W., Khan, T. M., Ab Mutalib, N. S., Goh, B. H., & Lee, L. H. (2019). Diversity of *Streptomyces* spp. from mangrove forest of Sarawak (Malaysia) and screening of their antioxidant and cytotoxic activities. *Scientific Reports, 9*(1). https://doi.org/10.1038/s41598-019-51622-x

Lee, D. H., Choi, S. L., Rha, E., Kim, S. J., Yeom, S. J., Moon, J. H., & Lee, S. G. (2015). A novel psychrophilic alkaline phosphatase from the metagenome of tidal flat sediments. BMC biotechnology, 15(1), 1. https://doi.org/10.1186/s12896-015-0115-2

Marziah, Z., Mahdzir, A., Musa, Md. N., Jaafar, A. B., Azhim, A., & Hara, H. (2016). Abundance of sulfur-degrading bacteria in a benthic bacterial community of shallow sea sediment in the off-Terengganu coast of the South China Sea. *MicrobiologyOpen, 5*(6), 967–978. https://doi.org/10.1002/mbo3.380

Mason, O. U., Hazen, T. C., Borglin, S., Chain, P. S. G., Dubinsky, E. A., Fortney, J. L., Han, J., Holman, H.-Y. N., Hultman, J., Lamendella, R., Mackelprang, R., Malfatti, S., Tom, L. M., Tringe, S. G., Woyke, T., Zhou, J., Rubin, E. M., & Jansson, J. K. (2012). Metagenome, metatranscriptome and single-cell sequencing reveal microbial response to Deepwater Horizon oil spill. *The ISME Journal, 6*(9), 1715–1727. https://doi.org/10.1038/ismej.2012.59

Mazalan, N., Zain, M. M., & Hamzah, A. S. (2012). Antimicrobial activity of marine bacteria from Malaysian coastal area. *2012 IEEE Symposium on Humanities, Science and Engineering Research*, 1273–1277. https://doi.org/10.1109/SHUSER.2012.6268808

Mazlan, A. G., Zaidi, C. C., Wan-Lotfi, W. M., & Othman, H. R. (2005). On the current status of coastal marine biodiversity in Malaysia. In *Indian Journal of Marine Sciences* (Vol. 34, Issue 1).

McKew, B. A., Taylor, J. D., McGenity, T. J., & Underwood, G. J. C. (2011). Resistance and resilience of benthic biofilm communities from a temperate saltmarsh to desiccation and rewetting. *The ISME Journal*, *5*(1), 30–41. https://doi.org/10.1038/ismej.2010.91

Ministry of Fisheries and Coastal Affairs, (Norway). (2009). *Marine bioprospecting – a source of new and sustainable wealth growth*. https://www.regjeringen.no/en/dokumenter/marine-bioprospecting--a-source-of-new-a/id575822/

Mohamad, N. I., Adrian, T. G. S., Tan, W. S., Muhamad Yunos, N. Y., Tan, P. W., Yin, W. F., & Chan, K. G. (2016). *Vibrio variabilis* T01: A tropical marine bacterium exhibiting unique N-acyl homoserine lactone production. *Frontiers in Life Science*, *9*(1), 17–23. https://doi.org/10.1080/21553769.2015.1066716

Mohamad, N. I., How, K. Y., Yin, W.-F., & Chan, K.-G. (2017). Whole-genome Sequencing of *Vibrio sinaloensis* T47, a Tropical Marine Isolate with Quorum Sensing Properties. *Journal of Genomics*, *5*, 48–50. https://doi.org/10.7150/jgen.16163

Morgan, X. C., & Huttenhower, C. (2012). Chapter 12: Human Microbiome Analysis. *PLoS Computational Biology*, *8*(12), e1002808. https://doi.org/10.1371/journal.pcbi.1002808

Mossop, J. (2015). *"Marine Bioprospecting" in The Oxford Handbook of the Law of the Sea* (D. Rothwell, A. O. Elferink, K. Scott, & Stephens Tim, Eds.). Oxford University Press.

Moulton, O. M., Altabet, M. A., Beman, J. M., Deegan, L. A., Lloret, J., Lyons, M. K., Nelson, J. A., & Pfister, C. A. (2016). Microbial associations with macrobiota in coastal ecosystems: patterns and implications for nitrogen cycling. *Frontiers in Ecology and the Environment*, *14*(4), 200–208. https://doi.org/10.1002/fee.1262

MYBIS, M. B. I. S. (2015). *Marine and coastal biodiversity*. https://www.mybis.gov.my/art/6

Naresh, S., Kunasundari, B., Gunny, A. A. N., Teoh, Y. P., Shuit, S. H., Ng, Q. H., & Hoo, P. Y. (2019). Isolation and partial characterisation of thermophilic cellulolytic bacteria from north Malaysian tropical mangrove soil. *Tropical Life Sciences Research*, *30*(1), 123–147. https://doi.org/10.21315/tlsr2019.30.1.8

Omar, S. M., Farouk, N. M., Malek, N. A., & Abidin, Z. A. Z. (2017). *Verrucosispora* sp. K2-04, Potential Xylanase Producer from Kuantan Mangrove Forest Sediment. *International Journal of Food Engineering*. https://doi.org/10.18178/ijfe.3.2.165-168

Popovic, A., Hai, T., Tchigvintsev, A. et al. (2017). Activity screening of environmental metagenomic libraries reveals novel carboxylesterase families. Sci Rep 7, 44103

Prakash, O., Shouche, Y., Jangid, K., & Kostka, J. E. (2013). Microbial cultivation and the role of microbial resource centers in the omics era. *Applied Microbiology and Biotechnology*, *97*(1), 51–62. https://doi.org/10.1007/s00253-012-4533-y

Priya, G., Lau, N.-S., Furusawa, G., Dinesh, B., Foong, S. Y., & Amirul, A.-A. A. (2018). Metagenomic insights into the phylogenetic and functional profiles of soil microbiome from a managed mangrove in Malaysia. *Agri Gene*, *9*, 5–15. https://doi.org/10.1016/j.aggene.2018.07.001

Rausch, P., Rühlemann, M., Hermes, B. M., Doms, S., Dagan, T., Dierking, K., Domin, H., Fraune, S., von Frieling, J., Hentschel, U., Heinsen, F. A., Höppner, M., Jahn, M. T., Jaspers, C., Kissoyan, K. A. B., Langfeldt, D., Rehman, A., Reusch, T. B. H., Roeder, T., … Baines, J. F. (2019). Comparative analysis of amplicon and metagenomic sequencing methods reveals key features in the evolution of animal metaorganisms. *Microbiome*, *7*(1). https://doi.org/10.1186/s40168-019-0743-1

Ser, H. L., Palanisamy, U. D., Yin, W. F., Abd Malek, S. N., Chan, K. G., Goh, B. H., & Lee, L. H. (2015). Presence of antioxidative agent, Pyrrolo[1,2-a] pyrazine-1,4-dione, hexahydro- in newly isolated *Streptomyces mangrovisoli* sp. nov. *Frontiers in Microbiology*, *6*(AUG). https://doi.org/10.3389/fmicb.2015.00854

Ser, H. L., Tan, W. S., Ab Mutalib, N. S., Yin, W. F., Chan, K. G., Goh, B. H., & Lee, L. H. (2016). Draft genome sequence of mangrove-derived *Streptomyces* sp. MUSC 125 with antioxidant potential. *Frontiers in Microbiology*, *7*(SEP). https://doi.org/10.3389/fmicb.2016.01470

Sien Aun Sia, E., Zhu, Z., Zhang, J., Cheah, W., Jiang, S., Holt Jang, F., Mujahid, A., Shiah, F. K., & Müller, M. (2019). Biogeographical distribution of microbial communities along the Rajang River-

South China Sea continuum. *Biogeosciences*, *16*(21), 4243–4260. https://doi.org/10.5194/bg-16-4243-2019

Simon, C., & Daniel, R. (2011). Metagenomic Analyses: Past and Future Trends. *Applied and Environmental Microbiology*, *77*(4), 1153–1161. https://doi.org/10.1128/AEM.02345-10

Song, J., Mujahid, A., Lim, P.-T., Samah, A. A., Quack, B., Pfeilsticker, K., Tang, S.-L., Ivanova, E., & Müller, M. (2017). Shotgun metagenomic analysis of microbial communities in the surface waters of the Eastern South China Sea. *Malaysian Journal of Microbiology*, *13*(4), 350–362. http://metagenomics.anl.gov/

Staley, J. T., & Konopka, A. (1985). Measurement of in Situ Activities of Nonphotosynthetic Microorganisms in Aquatic and Terrestrial Habitats. *Annual Review of Microbiology*, *39*(1), 321–346. https://doi.org/10.1146/annurev.mi.39.100185.001541

Thomas, T., Gilbert, J., & Meyer, F. (2012). Metagenomics-a guide from sampling to data analysis. *Microbial Informatics and Experimentation*, *2*(1), 3.

Vogel, M. A., Mason, O. U., & Miller, T. E. (2020). Host and environmental determinants of microbial community structure in the marine phyllosphere. *PloS One*, *15*(7), e0235441. https://doi.org/10.1371/journal.pone.0235441

Yasim, N. H. M. (2018). Isolation, identification and characterisation of lignocellulyic bacteria from mangrove roots.

Zainal Abidin, Z. A., Abdul Malek, N., Zainuddin, Z., & Chowdhury, A. J. K. (2016). Selective isolation and antagonistic activity of actinomycetes from mangrove forest of Pahang, Malaysia. *Frontiers in Life Science*, *9*(1), 24–31. https://doi.org/10.1080/21553769.2015.1051244

Open Water Integrated Multi-Trophic Aquaculture (IMTA) in Coastal Ecosystem: The Status and Prospects in Malaysia

Najiah, M.[1]*, Lee, K. L.[1], Nadirah, M.[1], Jalal, K. C. A.[2], Laith, A. A.[1], Habib, A.[1], Sheikh, H.I.[1], N.W. Rasdi[1], Zainathan, S.C.[1], Abu Hena, M. K.[1], Ruhil H. H.[3]

[1]*Faculty of Fisheries and Food Science, Universiti Malaysia Terengganu (UMT), 21030 Kuala Nerus, Terengganu*

[2]*Kulliyyah of Science, International Islamic University Malaysia (IIUM), Jalan Sultan Ahmad Shah, Bandar Indera Mahkota, 25200 Kuantan, Pahang*

[3]*Department of Paraclinical, Faculty of Veterinary Medicine, Universiti Malaysia Kelantan (UMK), Pengkalan Chepa, 16100 Kota Bharu, Kelantan*

Corresponding author: najiah@umt.edu.my

ABSTRACT

Globally, fish is an important source of affordable animal protein for humans. Amid the growing demand for seafood, aquaculture plays an important role to fill the supply shortfall of the stagnated capture fisheries to cater to the needs of the rising population. Malaysia's marine cage culture is confined to sheltered coastal water due to the constraints of low technological inputs. The single-trophic intensive cage culture is increasingly facing sudden massive fish death due to coastal pollution resulting from land-based anthropogenic activities and cage culture operation itself. Integrated multi-trophic aquaculture (IMTA) combines farming of different trophic species in proximity for symbiotic and complementary functions to foster ecological resilience, harmony and sustainability, as well as to help reduce diseases. Despite its infancy, IMTA has good prospects in bio-mitigating coastal pollution, restoring and preserving the vulnerable coastal ecosystems in Malaysia. There is no one-size-for-all IMTA system. An optimum species combination needs to be empirically determined based on the local economic and ecological scenarios.

Keywords: Marine Cage Culture, Self-pollution, Environmental Impacts, Bio-mitigation, Sustainability

INTRODUCTION

The present 7.7 billion world population is expected to rise to 9.7 billion by 2050 (United Nations, Department of Economic and Social Affairs, Population Division, 2019). The rising population is posing tremendous pressure and challenges to food and nutrition security, with over 820 million people in the world still suffering from hunger. Fish is an important source of affordable animal protein for humans, reaching 50 % of total intake or more in many least developed countries including those in the Asian region (FAO, 2020). As global capture fisheries stagnate in volume and increasingly fall short of growing world demand for seafood, the hope is on the continuously growing aquaculture to cater to the increasing demand (Figure 1). Endowed with a long coastline, Malaysia has a vast coastal front with potential sheltered waters for marine cage culture. The coastal cage farming is intensively operated almost entirely at a single trophic level, where different monospecies are cultured independently in different cages or areas. This single-trophic practice, over time, has led to pollution and degradation of the coastal environment, resulting in episodes of sudden massive death in cultured fish. This review discussed the status of open water IMTA in Malaysia, and its prospects in bio-mitigation of coastal pollution, restoration and preservation of vulnerable coastal ecosystems for sustainable development of marine cage culture.

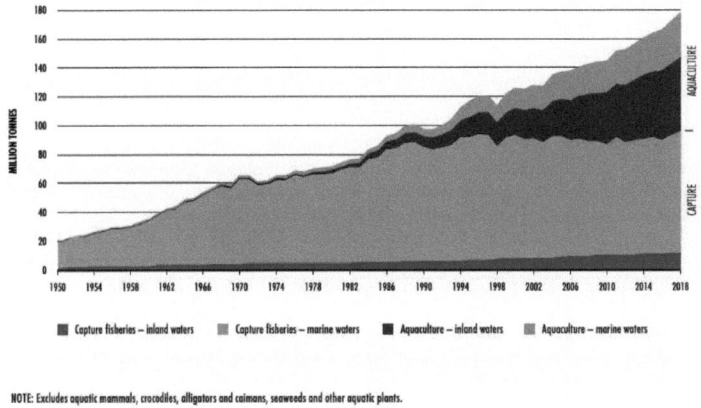

Fig. 1: World capture fisheries and aquaculture production (FAO, 2020).

Marine cage culture in Malaysia

Cage culture was first commercially established in the 1980s (Shariff and Gopinath, 2000). The low-level technology has confined cage culture to coastal regions protected from strong waves such as areas sheltered by islands, lagoons and estuaries. In the north, the Penang state has 30,961 units of cages with an area of 638,082 m², follow by Perak (17,840 cages, 363,458.46 m²) and Kedah (8,818 cages, 135,582.19 m²). In the central region, Selangor has 17,961 cages with 313,972.95 m². In the south, Johore has the most number of cages (8,856) with an area of 624,270 m². On the east coast, cage farming is predominantly located in Kelantan (5,622 cages, 57,283.88 m²) and Terengganu (2,047 cages, 40,956.82 m²). In the east Malaysia, Sabah and Sarawak has 8,699 cages (220,504 m²) and 1,630 cages (16,795 m²), respectively (DOF, 2018). The cage farming is almost solely single-trophic in practice, culturing finfish such as seabass, groupers and snappers, while a very small number of fish farmers are also doing line culture of organic extractive species, which depends on the availability of natural seed within the vicinity of the cage site. The single-trophic practice is increasingly facing tough challenges from sudden massive fish death due to declining quality of coastal waters.

Environmental and disease issues in marine cage culture

Marine cage culture can help relieve the fishing pressure on wild fish stocks, but if not managed, it can indeed be damaging to the ecosystem. Intensive cage farming can cause significant water quality deterioration due to feed waste and fecal inputs. It is estimated that 52 – 95 % of nitrogen (N) added to the culture system as food would eventually pollute the environment (Handy and Poxton, 1993), due to wastage, poor absorption and retention. Organic discharge from cage culture will deplete the dissolved oxygen (DO) in the water column through microbial degradation process (Hargrave et al., 1993). Also, microbial composting activity can directly cause high biochemical oxygen demand (Suratman et al., 2009). Besides, this process also increases the carbon dioxide production in the waterbodies due to respiration, and leads to low pH values. The self-pollution of cage farming, if unchecked, can cause eutrophication of waterbodies and seabeds, and induces excessive growth of algae and plants.

Furthermore, the coastal ecosystem is continuously exposed to anthropogenic contamination resulting from urbanization, industrialization and other economic activities. In a 10-year water quality surveillance (2003 to 2010 and 2014 to 2015) at mariculture site in Setiu Wetland Lagoon, Terengganu, Poh et al. (2019) revealed high phosphorus concentration related to oil palm plantation, high suspended solids due to large-scale land clearing, and ammonium enrichment resulting from land-based aquaculture discharge.

Aquaculture and anthropogenic pollution are continuously loading the coastal waters with a high amount of organic and inorganic wastes. Such waste substances not only trouble fish with depleted DO, ammonia poisoning and harmful algal bloom, but also predispose the cultured species to various disease agents (Najiah et al., 2002; Najiah et al., 2008; Ariff et al., 2019). In Malaysia, sudden massive fish deaths related to deteriorated water quality are becoming more often across the major coastal cage farming areas, incurring very heavy losses to farmers (Lim, 2019, August 12; Audrey, 2020, June 4; Lo, 2020, June 5). In this regard, mitigation measures are necessary to remediate nutrient-rich waters and prevent them from worsening to an intolerable extent for fish. This, in turn, will support the sustainable development of coastal aquaculture.

Integrated multi-trophic aquaculture
Integrated multi-trophic aquaculture is the farming of aquaculture species of different food chain levels in proximity for complementary ecosystem functions, whereby the uneaten feed and waste of one species are used by the species of other levels. For example, in the marine ecosystem, fed aquaculture species (e.g. finfish) are integrated with organic extractive species (e.g. suspension and deposit feeders) and inorganic extractive species (e.g. seaweeds). Figure 2 shows the schematic design of the open water IMTA system.

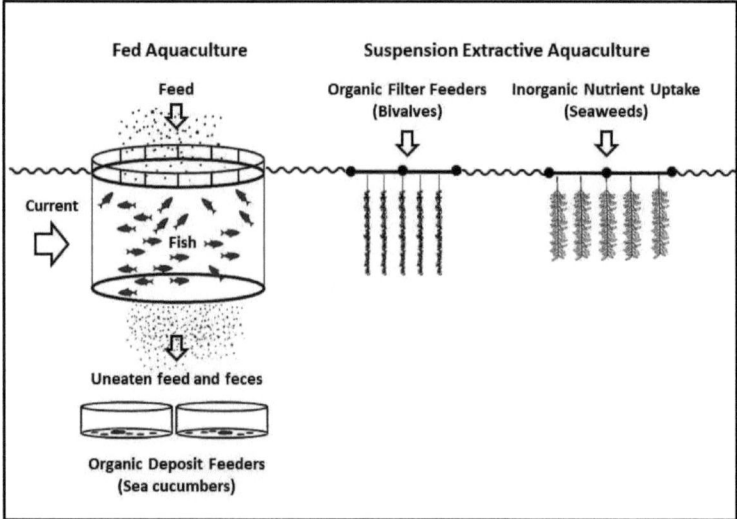

Fig. 2. Schematic view of an open water IMTA module showing the integration of fed aquaculture species (e.g. fish) with organic extractive species (e.g. bivalves as suspension filter feeders and sea cucumbers as deposit feeders) and inorganic extractive species (e.g. seaweeds). Deposit feeders are cultured below the fish cages to clean up uneaten feed and fish feces, while the filter feeders take up suspended organic particles, and the inorganic extractive species eliminates dissolved inorganic nutrients such as nitrogen and phosphorus.

The IMTA system has a long history in China involving bivalves and seaweeds. It has been successfully practised in Sanggou Bay since the late 1980s (Fang et al., 1996), and it is now widely applied in many parts of China. The abalone-kelp-sea cucumber combination is among the successful modules in practice. In Canada, the initial IMTA research took place in 2001 in the Bay of Fundy, on co-cultivation of salmon (*Salmo salar*), kelp (*Laminaria saccharina* and *Alaria esculenta*) and blue mussel (*Mytilus edulis*) (Chopin et al., 2007; Chopin and Robinson, 2004). The study showed increased growth in kelp and mussels by 46% and 50%, respectively, indicating an increase in food availability near the salmon farms. Chopin et al. (2007) also showed that, with proper management, the mussels and seaweeds produced from IMTA can be safely used for human consumption. Other countries that have also been exploring IMTA are Chile, South Africa and Israel (Chopin et al., 2008; Barrington et al., 2009), and more recently the United Kingdom (especially in Scotland), Ireland, Spain, Portugal, France, Turkey, Norway, Japan, Korea, Thailand, the U.S.A. and Mexico.(Garcia, 2012)

The IMTA approach is intended to reduce the environmental impacts of organic and inorganic wastes from aquaculture so that it can be more ecologically sustainable (Lefebvre et al., 2000; Chopin et al., 2008; Troell et al., 2003; Neori et al., 2017). It is considered a specialized form of the age-old polyculture practice that co-cultured various species in the water bodies, often without regard to the trophic level. From the economic perspective, IMTA is also a way to reduce economic risk, and to increase competitiveness through species diversification (Barrington et al., 2009). It is increasingly gaining importance for its yield quality and environmental compatibility. Table 1 shows some of the experimental IMTA modules in Southeast Asia.

Table 1: Experimental IMTA modules in some Southeast Asian countries.

Country	Species combination	Results	Reference
Gerupuk Bay, Central Lombok, Indonesia,	Tiger grouper (*Epinephelus fuscoguttatus*), silver pompano (*Trachinotus blochii*) and seaweed (*Kappaphycus alvarezii*)	Good growth performance in both grouper and pompano, and increased seaweed production	Radiarta and Erlania, 2016
Gerupuk Bay, Central Lombok, Indonesia,	(*Eucheuma cottonii* - lobster - abalone); (*E. cottonii* - abalone - red carp); (*E. cottonii* -abalone - grouper); (*E. cottonii* - abalone - pomfret)	*E. cottonii* - abalone - grouper combination showed the highest biomass production of *E. cottonii*	Sukiman et al., 2014.
Southern Cebu, Philippines	Donkey's ear abalone (*Haliotis asinine*) as fed species and seaweeds (*Gracilaria heteroclada* and *Eucheuma denticulatum* as inorganic extractive species	Abalone culture did not produce a high amount of wastes at the experimental farming scale. *Gracilaria* and *Eucheuma* grown side by side abalone cages serve as feed-on-demand and biofilters for inorganic wastes	Largo et al., 2016
Guimaras, Philippines	Combined pen culture of milkfish *Chanos chanos*, with sea cucumber *Holothuria scabra* and seaweed *Kappaphycus* sp.	Mitigated the impacts of excess nutrients from uneaten feed and feces of milkfish, and obtained additional income from non-fed species	SEAFDEC, 2017
Khánh Hòa province, Vietnam	Sea cucumber with shrimp or babylon snails	Low-cost culture of sea cucumber improved water quality for shrimp or babylon snails	The Fish Site, 2019

| Sabah, Malaysia, | Spiny lobster (*Panulirus ornatus*), sea cucumber (*Holothuria scabra*) and seaweed (*Kappaphycus alvarezii*) in recirculating system and flow-through system | Better efficiency of water quality remediation and growth in flow-through system | Sumbing et al., 2016 |

Status and prospects of IMTA in Malaysia

The IMTA concept is in its infancy in Malaysia at present. In Terengganu and Kelantan, depending on the availability of wild seeds, some cage cultures do practise dropper-line culture of oyster alongside seabass or grouper culture for additional income rather than from the ecological perspective. In this regard, ecological awareness education and technical support will help the farmers adopt the complete IMTA module. Blessed with an extensive coastline and numerous islands, Malaysia has various habitats for a good variety of seaweeds with 35 species in 12 families of Cyanophyta; 113 species in 16 families of Chlorophyta; 95 species in 8 families of Ochrophyta; and 216 species in 36 families of Rhodophyta. Despite having rich resources of seaweeds, so far only *Kappaphycus alvarezii*, *Eucheuma denticulatum* and *Gracilaria manilaensis* are identified suitable for commercial purposes (Phang et al., 2019). Seaweeds are now most widely cultivated in Sabah with 9,835.30 Ha of farming areas, while Kedah has very small-scale cultivation of 0.68 Ha (DOF, 2018). With much-established seaweed cultivation, and 220,504 m² (8,699 cages) of cage culture, Sabah may have a better opportunity of implementing IMTA compared with other states.

CONCLUSION

Malaysia's marine cage culture is at a crossroads as the pollution from land-based anthropogenic activities and cage farming itself continuously disrupts the ecosystem homeostasis. It may not be too long before the pollution-related massive fish death becomes overwhelmingly heavy, and makes the farming operation not commercially viable. Despite its infancy in Malaysia, IMTA has good prospects for bio-mitigation of coastal pollution, and to restore and preserve the vulnerable coastal ecosystem. The symbiotic and complementary nature of IMTA will promote ecological resilience, harmony and sustainability, as well as reduce disease probability in cultured species. Nevertheless, there is no one-size-fits-all IMTA system. A successful module in a locality is unlikely to fit in all places. The optimum species combination should be empirically determined based on the local economic and ecological scenarios.

REFERENCES

Ariff, N., Abdullah, A., Azmai M.N.A., Musa N., & Zainathan, S.C. (2019). Risk factors associated with viral nervous necrosis in hybrid groupers in Malaysia and the high similarity of its causative agent nervous necrosis virus to reassortant red-spotted grouper nervous necrosis virus/striped jack nervous necrosis virus strains. *Veterinary World*, 12(8), 1273-1284.

Audrey, D. (2020, June 4). No need to worry about fish carcasses at sea. *New Straits Times*. Retrieved from https://www.nst.com.my/news/nation/2020/06/597957/no-need-worry-about-fish-carcasses-sea.

Barrington, K., Chopin, T., & Robinson, S. (2009). Integrated multi-trophic aquaculture (IMTA) in marine temperate waters. In D. Soto (ed.). Integrated mariculture: a global review. *FAO Fisheries and Aquaculture Technical Paper*. No. 529. Rome, FAO. pp. 7–46.

Chopin, T., & Robinson, S. (2004) Defining the appropriate regulatory and policy framework for the development of integrated multi-trophic aquaculture practices:introduction to the workshop and positioning of the issues. *Bull Aquacult Assoc Can.*, 104, 4–10.

Chopin, T., Robinson, S., Page, F., Ridler, N., Sawhney, M., Szemerda, M., Sewuster, J., & Boyne-Travis, S. (2007). Integrated multi-trophic aquaculture making headway in Canada. *The Canadian Aquaculture Research and Development Review*, p. 28.

Chopin, T., Robinson, S.M.C., Troell, M., Neori, A., Buschmann, A.H., & Fang, J. (2008). Multitrophic Integration for Sustainable Marine Aquaculture. In Sven Erik Jørgensen and Brian D. Fath (Editor-in-Chief), *Ecological Engineering*. Vol. [3] of *Encyclopedia of Ecology*, 5 vols. pp. 2463-2475. Oxford: Elsevier.

DOF. (2018). Annual Fisheries Statistics. Retrieved from https://www.dof.gov.my/dof2/resources/user_29/Documents/Perangkaan%20Perikanan/2018%20Jilid%201/Table_akua_2018_-new.pdf

Fang, J., Kuang, S., Sun, H., Li, F., Zhang, A., Wang, X., & Tang, T. (1996). Mariculture status and optimizing measurements for the culture of scallop *Chlamys farreri* and kelp *Laminaria japonica* in Sanggou Bay. *Mar Fish Res*, 17, 95-102.

FAO. (2020). The State of World Fisheries and Aquaculture 2020. Sustainability in action. Rome. https://doi.org/10.4060/ca9229en

Garcia, J. (2012). Sustainable alternative for diversifying cultures and for protecting marine environment quality. In Integrated Multi-trophic Aquaculture (IMTA): A sustainable, pioneering alternative for marine cultures in Galicia (ed. Guerrero, S. and Cremades, J.), pp. 9. Regional Government of Galicia (Spain), Regional Council of the Rural and Regional Maritime Environment Marine Research Centre, Spain. https://hal.archives-ouvertes.fr/h

Handy, R.D., & Poxton, M.G. (1993). Nitrogen pollution in mariculture: toxicity and excretion of nitrogenous compounds by marine fish. *Rev. Fish. Biol. Fisheries*, 3, 205-241.

Hargrave, B.T., Duplisea, D.E., Pfeiffer, E., & Wildfish, D.J. (1993). Seasonal changes in benthic fluxes of dissolved oxygen and ammonium associated with marine cultured Atlantic salmon. *Marine Ecology Progress Series*, 96, 249-257.

Largo, D.B., Diola, A.G., & Marababol, M.S. (2016). Development of an integrated multi-trophic aquaculture (IMTA) system for tropical marine species in Southern Cebu, Central Philippines. *Aquaculture Reports*, 3, 67-76.

Lefebrve S., Barille', L., & Clerc, M. (2000). Pacific oyster (*Crassostrea gigas*) feeding responses to a fish farm effluent. *Aquaculture*, 187, 185-198.

Lim, C. (2019, August 12). Fish breeders hit badly again as 50, 000 fishes found dead in Teluk Bahang. *The Star*. Retrieved from https://www.thestar.com.my/news/nation/2019/08/ 12/fish-breeders-hit-badly-again-as-50-000-fishes-found-dead-in-teluk-bahang

Lo, T.C. (2020, June 5). Red tide heading towards Kedah. *The Star*. Retrieved from https://www.thestar.com.my/news/nation/2020/06/05/killer-red-tide-heading-towards-kedah

Najiah, M., Lee, K.L., Hassan, M.D., Muhd-Azmi, M. L., & Shariff, M. (2002). Morphological, biochemical and physiological characteristics of *Vibrio parahemolyticus* isolates in diseased fish and shrimp ponds in Malaysia. *Jurnal Veterinar Malaysia*, 14(1&2), 25-30.

Najiah, M., Nadirah, M., Lee, K. L., Lee, S.W, Wendy, W., Ruhil, H.H., & Nurul, F.A. (2008). Bacterial flora and heavy metals in cultivated oysters *Crassostrea iredalei* of Setiu Wetland, East Coast Peninsular Malaysia. *Veterinary Research Communication*, 32, 377-381.

Neori, A., Shpigel, M., Guttman, L., & Israel, A. (2017). Development of polyculture and integrated multi-trophic aquaculture (IMTA) in Israel: a review. *The Israeli Journal of Aquaculture-Bamidgeh*, 69:1-19.

Phang, S.M., Yeong, H.Y., & Lim, P.E. (2019). The seaweed resources of Malaysia. *Botanica Marina*, 62(3). https://doi.org/10.1515/bot-2018-0067

Poh, S. C., Ng, N.C.W., Suratman, S., Mathew, D., & Mohd Tahir, N. (2019). Nutrient availability in the Setiu Wetland Lagoon, Malaysia: trends, possible causes and environmental impacts. *Environmental Monitoring and Assessment*, 191, 3. https://doi.org/10.1007/s10661-018-7128-y

Radiarta, N., & Erlania. (2016). Performance of mariculture commodities under Integrated Multi-Trophic Aquaculture (IMTA) system at Gerupuk Bay, Central Lombok, West Nusa Tenggara. *Jurnal Riset Akuakultur*, 11 (1), 85-97.

SEAFDEC. (2017). Southeast Asian State of Fisheries and Aquaculture. Southeast Asian Fisheries Development Center, Bangkok, Thailand. 167 pp.

http://repository.seafdec.org/bitstream/handle/20.500.12066/6204/6.5-Addressing-concerns-due-to-aquaculture-climate-change.pdf?sequence=1&isAllowed=y

Shariff, M., & Gopinath, N. (2000). Cage culture in Malaysia: an overview [Paper presentation]. In *Cage Aquaculture in Asia*: Proceedings of the First International Symposium on Cage Aquaculture in Asia (pp. 75-81). Asian Fisheries Society, Manila, and World Aquaculture Society - Southeast Asian Chapter, Bangkok.

Sukiman, Faturrahman, Rohyani I.S., & Ahyadi, H. (2014). Growth of seaweed *Eucheuma cottonii* in multi trophic sea farming systems at Gerupuk Bay, Central Lombok, Indonesia Nusantara. *Bioscience*, 6, 82-85.

Sumbing, M.V., Al-Azad, S., Estim, A., & Mustafa, S. (2016). Growth performance of spiny lobster *Panulirus ornatus* in land-based Integrated Multi-Trophic Aquaculture (IMTA) system. *Transactions on Science and Technology*, 3(1-2), 143-149.

Suratman, S., Awang, M., Loh, A.L., & Mohd Tahir, N. (2009). Water quality index study in Paka River basin, Terengganu (in Malay). *Sains Malaysiana*, 38, 125-131.

The Fish Site. (2019). Vietnam promotes sea cucumber IMTA. Retrieved from https://thefishsite.com/articles/vietnam-promotes-sea-cucumber-imta

Troell, M., Halling, C., Neori, A., Chopin, T., Buschman, A.H., Kautsky, N., & Yarish, C. (2003). Intergrated mariculture: asking the right questions. *Aquaculture*, 226, 69-90.

United Nations, Department of Economic and Social Affairs, Population Division. (2019). World Population Prospects 2019: Highlights (ST/ESA/SER.A/423).

Antioxidant Properties of *Nerita articulata* from Estuarine Mangrove Kuantan, Pahang Malaysia

Deny Susanti[1*], Mohd Faizol, A.L[2]

[1]*Department of Chemistry, Kulliyyah of Science, International Islamic University Malaysia, 25200 Kuantan, Pahang, Malaysia.*
[2]*Department of Biotechnology, Kulliyyah of Science, International Islamic University Malaysia, 25200 Kuantan, Pahang, Malaysia.*
Corresponding author: deny@iium.edu.my.

ABSTRACT
Mollusks are one of the main macroinvertebrates that play a significant ecological role in nutrient dynamics in the mangrove ecosystem because they form an essential link within the food web as predators, herbivores, detritivores and filter feeders. They are useful bioindicators of environmental pollution, due to their methods of filter-feeding. Based on the above contexts, the antioxidant properties of mollusc species *Nerita articulata* was investigated in a mangrove estuary, Kuantan, Pahang around the east coast of Malaysia. In the present study, different antioxidant tests were conducted to evaluate the antioxidant activities of water, methanol and dichloromethane: methanol extracts of *N. articulata*. The results were compared with alpha-tocopherol and ascorbic acid, which are generally known as antioxidant compounds. The percentage of scavenging activities and lipid peroxidation inhibition for each of the extracts were also determined. The extracts were found to have different levels of antioxidant properties in the test models used. All extracts had strongly inhibited lipid peroxidation and also had shown low radical scavenging activities. Therefore, this species could be considered as a significant antioxidant source in terms of lipid peroxidation. The study indicates that these extracts from the mollusc *N. articulata* have good antioxidant activities that can be harnessed as leads for potential bioactive compounds.

Keywords: *Nerita articulate*, Antioxidant activity, Free radical, Scavenging activity, Lipid peroxidation.

INTRODUCTION
Marine or natural aquatic products have attracted the attention of biologists and chemists the world over for the past five decades. As a result of the potential for new drug discovery, natural aquatic products have attracted scientists who led to the discovery of thousands of aquatic-based products to date, and many of the compounds have shown promising biological activity. The biological activities of an extract of marine organisms or isolated compounds are categorised in terms of antimicrobial, antileishmanial, anthelmintic, antimalarial, anti-inflammatory, antioxidant, anticancer and antiallergic activity (Anand, 2010; Malve, 2016). Mollusks are considered as one of the important sources to derive bioactive compounds that exhibit antitumor, antimicrobial, anti-inflammatory and antioxidant activities (Sole et al., 1994; Bhakuni and Rawat, 2005; Benkendorff et al., 2010). Mollusks also contain rich nutrients that are beneficial to people of all ages. In our body, the oxidation process leads to cell damage, cancer and degenerative diseases; antioxidant molecules present in different mollusks prevent cell damage from oxidation reaction (Nagash et al., 2010). Compounds isolated from mollusks were also used in the treatment of rheumatoid arthritis and osteoarthritis (Chellaram and Edward, 2009). Molluscan extracts also exhibited antiviral and antibacterial activity against pathogenic fish bacteria, and the extract also may be applied in aquaculture (Defer et al., 2009).

Mangroves are documented to be among the world's most productive ecosystems which provide important nursery and feeding grounds for juvenile fish and potential invertebrate species such as mollusks (Siraprapha et al., 2016). Mollusks are one of the main macroinvertebrates that play a significant ecological role in nutrient dynamics in the mangrove ecosystem because they form an important link within the food web as predators, herbivores, detritivores and filter feeders. They are useful bio-indicators of environmental pollution, due to their methods of filter-feeding. *N. articulata* is most dominant and inhabits widely in the mangrove area of Kuantan estuary.

Based on the above perspectives, this study was conducted to observe the antioxidant properties of the selected dominant mollusc species which have been found abundantly near the Kuantan estuarine mangrove area. The study was aimed to determine antioxidant activities of *Nerita's* crude extracts using different techniques (free radical or lipid peroxidation) and to analyse the quantitative aspects of antioxidant activities in selected mollusc species.

METHODOLOGY
Sampling Area
Kuantan mangrove area located near the estuarine region of Kuantan river with latitude 3° 48' 20.63 °N and latitude 103° 20' 3.36 °E. It is under the district of Kuantan about 2 kilometres away from Kuantan city. The area was surrounded by the 339 hectares of Mangrove Reserve Forest which had been in existence for over 500 years. This study area is recognised as the habitat for a variety of animals like birds, fish and other potential invertebrates like gastropods, arthropods.

Fig 1: Sampling Area: Mangrove Estuary Kuantan

Sample Collection
The fresh samples of *Nerita* species were collected from the estuarine mangrove area, Kuantan. The samples were kept in a plastic bag before it was stored in a cold room. After that, the body and the shell were separated, and the study of antioxidant properties was focused on the body part of *Nerita* sp. Then the samples were stored at -20 °C until extraction (Houssen and Jaspars, 2005; Bhakuni and Rawat, 2005). The

species were identified until the genus level and referred to the Taxonomy and Distribution of Neritidae (Mollusc: Gastropoda) in Singapore discussed by Siong and Reuben, 2008; Bouchet and Rocroi, 2005).

Extraction by Different Solvents
The samples were extracted as a function of their polarity using water and organic solvents. The solvents are water, dichloromethane (DCM): methanol and methanol extractions (Sies, 1997; Houssen and Jaspars, 2005; Bhakuni and Rawat, 2005). The details extraction methods were conducted by different solvents which are described as follows:

Water Extraction
The samples were cut into small pieces, and the weighed of the samples were recorded accordingly. Then the samples (304.37 g) were added with 500 mL distilled water and ground using a blender. The mixture was transferred in a conical flask and stored in a cold room (0 °C) for 24 hours. Later, the samples were filtered using Whatman No. 1 filter paper, and the residues/ filtrate were collected for organic solvent extraction. The aqueous extract was frozen in the deep freezer (-20 °C). Then the samples were freeze-dried, and the crude extract for aqueous extraction was obtained.

Dichloromethane: Methanol Extraction
The samples were weighed (432.78 g) and then soaked with 500 mL of DCM: methanol (1:1), blended and stored for 24 hours at room temperature. Then the soaked samples were filtered, and the residues/filtrate were collected for methanol extraction. The samples were dried up in fume chamber to remove remaining solvent for 1-3 days. The crude extract for DCM: methanol extraction was obtained and kept into the freezer.

Methanol Extraction
The samples were weighed (391.51 g) and added with 500 mL methanol and then blended. After that, the sample was kept for 24 hours at room temperature, filtered, and the extracts were evaporated accordingly. The samples were dried up in a fume chamber for 1-3 days. The crude extract for methanol extraction was obtained and kept in the freezer.

Antioxidant Screening
Rapid Screening using Dot-Blot and DPPH Staining
Rapid screening of antioxidants referred to the method Dot-Blot and DPPH staining with a slight modification to detect antioxidant properties in dried samples of the freezer. The crude extracts were dissolved with methanol with concentration 10mg/ml. The extracts and vitamin C were cautiously loaded on the TLC layer and dried for 3 minutes. Then 0.4 mM DPPH solution was sprayed onto the TLC layer. The stained TLC layer revealed a purple background with a white spot at the location of the drops, which showed radical scavenger capacity (Soler-Rivas et al., 2000; Subhapradha et al., 2013).

Antioxidant Quantitative Assay
Free Radical Scavenging Assay
The DPPH radical scavenging activity of the extracts from the samples *N. articulata* was determined by using the protocol of Brand-William el al. (1995). Cited By in Scopus (1464)The free radical scavenging activity of different extracts was assessed. The stock solution of each extract dissolving in methanol with 10 mg/mL concentration was prepared. The serial dilution was performed triplicates in 500, 250, 125, 62.5, 31.3, 15.6, 7.8 µg/mL in concentration from the stock solution. Each extract (100 µL) was mixed with 3.9 mL of a freshly prepared solution containing 25 mg/L of 1,1-diphenyl-2-picrylhydrazyl (DPPH) radicals in methanol. The absorbance was measured by UV light at 515 nm 30 min later. The percentage of DPPH scavenging activity was calculated as follows:

Scavenging activity (%) = [1-(absorbance of sample/absorbance of blank)] x 100

A lower absorbance indicates a higher scavenging effect. EC_{50} value (mg/mL) is the effective concentration at which DPPH radicals were scavenged by 50%. Vitamin C and E were used as positive controls.

Ferric Thiocyanate (FTC) Method
The FTC method was followed as adopted by Huang et al. (2005). This method was slightly modified in this study. 4 mg crude extract was dissolved in 4 mL of 95% (w/v) ethanol was mixed with linoleic acid (2.51%, v/v) in 99.5% (w/v) ethanol (4.1 mL), 8 mL of 0.05M phosphate buffer pH 7.0 and 3.9-mL distilled water. The mixture was stored in a screw-cap container at 40^0C in the dark. 0.1 mL of this mixture was added with 9.7 mL of 75% ethanol and 0.1 mL of 30% (w/v) ammonium thiocyanate. Precisely 3 minutes after the addition of 0.1 mL of 20 mM ferrous chloride in 3.5% (v/v) hydrochloric acid to the reaction mixture, the absorbance at 500 nm of the resulting red solution was measured. Then it was measured again every 24 hours of the following days when the absorbance of the control reached the maximum value. The percent inhibition of linoleic acid peroxidation was calculated as:

Inhibition (%) =100 - [(absorbance of increase of sample/absorbance increase of control) x 100]
All tests were run in triplicate and vitamin E as a positive control.

Descriptive Analysis
All experiments were performed in triplicates. The results were presented in mean ± standard deviation. This analysis was a descriptive statistic. As for the data and graphs; they were subjected to analyses using Microsoft® Office Excel 2007 and ANOVA.

RESULTS AND DISCUSSION
Sample Identification
This snail in a pin-striped suit was commonly seen in mangroves, often occurring in large numbers. It may also be seen on rocky shores, especially those near mangroves. Tan and Clements (2008) observed this snail on mangrove tree trunks and roots, monsoon canal walls, muddy banks, and rocky areas in or near mangroves. It was also known as *N. lineata*. The size of this species was 2-3 cm along with a shell which was sturdy and rounded. The colour of this species was beige, grey or pinkish with fine, spiralling black ribs. The flat underside of the shell was white, sometimes with yellow patches. There were small teeth at the shell opening. The Operculum was evenly covered with tiny bumps. The animal had fine black lines and long thin black tentacles. It grazed on algae and appeared to return to the same spot after a feeding bout. According to Tan & Clements (2008), the lined Nerite was probably the most widely distributed. This species was most abundant in monsoon canals, walls and mangrove trees, somehow numbering in the hundreds in a single location. The table below described the picture and morphology of the species. The most dominant gastropod in the Kuantan mangrove area was identified morphologically as follows (Table 1 and Figure 2):

Table 1: Taxonomy of *N. articulata*

Phylum	Mollusca
Class	Gastropoda
Order	Neritopsina
Family	Neritidae
Genus	*Nerita*
Species	*Nerita articulata*

N. articulata	Characteristics
	- Shell sturdy and rounded - Colour: beige, grey or pinkish with fine, spiralling black ribs - Size: 2-3cm
	- Flat underside the shell: white, sometimes with yellow patches - small teeth at the shell opening
	- Habitat: commonly seen on a mangrove; mangrove's tree trunks and roots, tree's hole, monsoon canal walls, muddy banks, rocky areas in or near mangroves

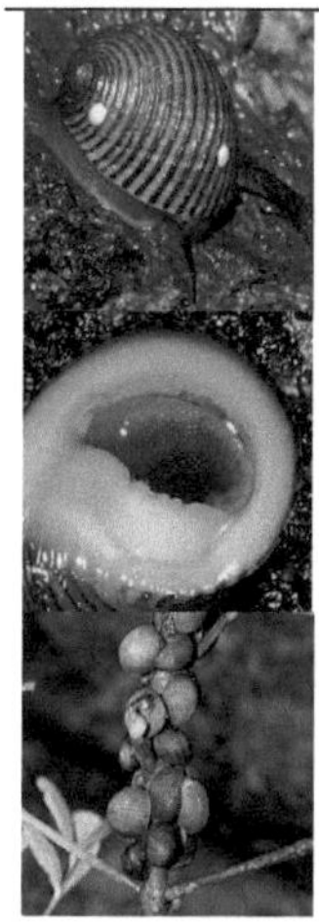

Fig. 2: The characteristics of the *N. articulata*

Sample Extraction

The selection of a suitable extraction procedure could increase the yield of antioxidant compounds relative to the plant material. Several extraction techniques have been patented using solvents with different polarity (such as petrol, ether, hexane, toluene, acetone, methanol and ethanol) as well as the assay techniques and substrate used (Mayer & Hamann, 2005). Bioactive compounds were extracted according to their polarity using water and organic solvents. The applied extraction methods were water extraction, dichloromethane (DCM): methanol extraction and methanol extraction (Houssen & Jaspars, 2005; Tinu et al., 2019).

Table 2: Weight of the extracted sample using different solvent

Method	Weight of mollusc body before extraction (g)	Weight of crude extraction after dried (g)	Yield (%)	Observation of extracts
Water Extraction	304.73	12.69	4.16	Light grey colour, powder form
Methanol Extraction	391.51	9.70	2.48	Dark brown colour, sticky form
Dichloromethane: Methanol Extraction	432.78	2.03	0.47	Dark green colour, sticky form

Technically, the solvent extracted the biological compound due to its polarity. Therefore, the different bioactive compound was extracted in each extraction. In Table 2, it showed the weight of the crude extract for each of the solvent. The biological compound was extracted the most by using water as the solvent. Water was generally known as the universal solvent. Based on the result, it might indicate that molecule constituents in the species were more soluble in the polar solvent. However, a solution that was extracted by the water did not mean to have the most antioxidant properties as it was only determined by three separated methods; dot-blot, scavenging activity method, and ferric thiocyanate (FTC).

Antioxidant Screening
Rapid Screening of Antioxidants using Dot-Blot and DPPH Staining
Rapid screening of antioxidants using the method Dot-Blot and DPPH staining was described by Soler-Rivas et al. (2000) with slight modification. Dot-Blot and DPPH staining was the first method that had been used to screen antioxidant properties in this study. Different types of solvent extraction with concentration 10 mg/ml was placed on the TLC plate, and the antioxidant properties were detected after DPPH staining. The appearance of white spots indicates the presence of antioxidants of different extract from samples in the dot blot (Huang et al., 2005). This method was based on the inhibition of the accumulation of oxidised compounds since the addition of antioxidants inhibited the generation of free radicals.

Vitamin C was used as the control for this experiment. All extracts showed a positive result, but they were slightly different in their intensity. The intensity of the white/yellow colour depended on the amount and nature of radical-scavenger present in the extract (Rahman et al., 2015). A white/yellow spot appeared in the methanol and DCM: Methanol extracts indicated these samples extracted high intensity of the antioxidant compounds. However, the low intensity of antioxidant compounds had been extracted by using water as solvent (Table 3).

Quantitative Test
Free Radical Scavenging Activity
This method is currently popular based upon the use of the stable free radical diphenylpicrylhydrazyl (DPPH). The purpose of this study was to evaluate the scavenging effects of the extracts of *N. articulata,* from different solvent extraction, know the basics of the method and also to understand the use of the parameter "EC_{50}" (equivalent concentration to give 50% effect) which was currently used in the interpretation of experimental data from the method.

2, 2-Diphenyl-1-picrylhydrazyl was characterised as a free radical under the delocalisation of the spare electron over the molecule as a whole, so that the molecules did not dimerise, as would be the case with most other free radicals. The delocalisation also gave rise to the deep violet colour, characterised by an absorption band in methanol solution centred at about 515 nm (Molyneux, 2004). When a solution of DPPH was mixed with that of a substance that can donate a hydrogen atom, then this gave rise to the reduced form with the loss of this violet colour. This condition indicated that DPPH radical was scavenged by antioxidants through the donation of hydrogen, forming the reduced DPPH-H.

Table 3: Antioxidant activities of different solvent extracts from *N. articulate*

Scavenging activity (%) = [1-(absorbance of sample/absorbance of blank)] x 100			
Concentration(µL)	Water Extract (±SD)	Methanol Extract (±SD)	DCM/Methanol Extract (±SD)
1000	4.4829 ± 0.013	6.5104 ± 0.044	7.6198 ± 0.085
500	4.2969 ± 0.008	4.9665 ± 0.017	5.2141 ± 0.028
250	3.7016 ± 0.005	4.9200 ± 0.007	3.0187 ± 0.008
125	4.3527 ± 0.005	5.0223 ± 0.024	2.4058 ± 0.024

62.5	4.1388 ± 0.01	6.9289 ± 0.038	8.9645 ± 0.044
31.3	4.5480 ± 0.008	5.6176 ± 0.021	6.1288 ± 0.055
15.6	4.3992 ± 0.01	8.2682 ± 0.059	5.4336 ± 0.044
7.8	9.0681 ± 0.063	7.4498 ± 0.036	4.2444 ± 0.022

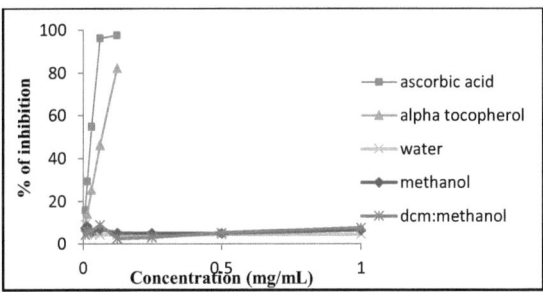

Fig. 3: Percentage of inhibition that shows IC$_{50}$ for ascorbic acid, alpha-tocopherol, water extract, methanol extract and dichloromethane: methanol extract.

Table 3 showed antioxidant activities of different solvent extracts from *N. articulata*. Sample extracts (10 mg), with various solvents, have reacted with DPPH free radical. All of the samples exhibited low antioxidant activity (2.4058–9.0681%). None of the samples exceeded 10% of antioxidant scavenging activities which indicated that the antioxidant scavenging activities of   *N. articulata* were deficient. Besides, the concentration for three samples cannot be determined since the graph in Figure 3 did not rise to 50% of the inhibition.

According to Manduzio et al. (2005) on oxidative stress in mollusks showed the level of malondialdehyde (MDA) increased from 4.48 ± 0.24 nmol/mg to 7.58 ± 0.38 nmol/mg after 168-hour anoxia. In the cell of the digestive gland, the level of MDA rose more than three times (from 2.7 ± 0.14 nmol/mg to 8.48 ± 0.43 nmol/mg). This statistic showed the level of lipid peroxidation was almost the same in *Nerita articulate.*

Numerous methods and modifications have been proposed to evaluate antioxidant activity and to explain how antioxidants function. Of these, the DPPH assay, the reducing ability, metal ion chelating, and active oxygen species quenching assay are most commonly used for the evaluation of antioxidant activities of extracts (Nadezhda, 2008). The absorption maximum of a stable DPPH radical in methanol was at 517 nm. The decrease in absorbance of the DPPH radical caused by antioxidants, because of the reaction between antioxidant molecules and radical, progresses, which results in the scavenging of the radical by hydrogen donation. It is visually noticeable as discolouration from purple to yellow.

Ferric Thiocyanate (FTC) Assay
Ferric thiocyanate (FTC) assay determined the amount of peroxide produced during the initial stages of oxidation which were the primary products of oxidation, and it was representing the *in vivo* condition. Compared to the DPPH assay, the DPPH free radical was synthetic radicals or *in vitro* radicals, which meant that it did not exist in the human body. This assay was significant because it represented what was happening in the human body. The crude extracts that showed antioxidant free radical scavenger character did not mean that it would function properly in the human body. There was a possibility that the compounds became pro-oxidant after scavenging the radicals. Pro-oxidant was the state where the antioxidant itself became free radical and directly caused the chain reaction propagation. If the crude inhibited high in this FTC assay, the compound could be considered safe to be consumed.

The reaction mixture of linoleic acid, ethanol, phosphate buffer and antioxidant (sample and standard) was incubated at 40 °C, and the peroxide value was measured by the absorbance at 500 nm after reaction between $FeCl_3$ and thiocyanate. In this test, the linoleic acid (RCOOH) was reduced by Fe^{2+} to free radical (RO⁺), while the ferrous ion itself undergoes the oxidation process to Fe^{3+}. Then, the Fe^{3+} ion reacts with thiocyanate ion (SCN)⁻ to give complex $Fe(SCN)_3$ as a bright red colour. Absorbance intensity of complex $Fe(SCN)_3$ was measured by spectrophotometer. The low absorbance values corresponding to a high percent of inhibition thus indicate that the sample could inhibit lipid peroxidation. The low absorbance values corresponding to a high percent of inhibition, therefore suggest that sample could inhibit lipid peroxidation (Deny et al., 2006).

The antioxidant effects of *Nerita* species extract and vitamin E on the peroxidation of linoleic acid were investigated, and the results were presented in Table 3 and Figure 4.

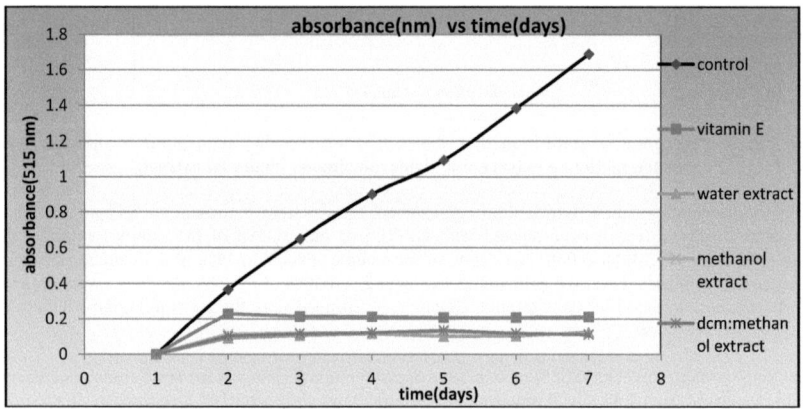

**Figure 4: Absorbance of extracts at 4 mg/mL concentration using the FTC method.
Results are of duplicate measurements**

The absorbance ranges recorded for sample, vitamin E and control were $0.0629 \pm 0.003 - 0.1269 \pm 0.001$, $0.000 - 2.113$ and $0.1692 \pm 0.001 - 0.2084 \pm 0.002$, respectively. From the graph showing the absorbance of all samples has increased by the time. The test was stopped after the reduction of absorbance occurred. The graph showed a strong lipid peroxidation inhibition by the *Nerita* sample extracts. The graph of the sample was below the vitamin E graph, meaning that the sample was stronger inhibition than the vitamin E. Moreover, all the crude extract inhibition percentage was close even below to the graph of vitamin E, which meant the samples contained strong lipid peroxidation inhibitor (Figure 4).

Each extract showed strong antioxidant activity in inhibition of linoleic acid peroxidation at a concentration of 4 mg/ml, as compared to the control ($p < 0.05$), and significantly prolonged the induction period of auto-oxidation of linoleic acid. From FTC results, the percentage inhibition of peroxidation in the linoleic acid system by 10 mg of water, methanol and DCM: methanol extracts were found to be $92.66 \pm 0.02\%$, $93.19 \pm 0.003\%$ and $93.4932 \pm 0.007\%$ respectively at the eight days of testing. These values were significantly ($p < 0.05$) higher than that exhibited by 1 mg of α-tocopherol (87.5%). A similar report by Xiu et al. (2019), found that the extract of mollusk, *Tergillarca granosa* strongly inhibits lipid peroxidation as well.

CONCLUSION

Based on the results from this study indicated that *N. articulata* has significant antioxidant activity. The data on extraction procedures and antioxidant activity assessment obtained from DCM: methanol, methanol and water extracts, suggested *N. articulata* is promising sources to isolate the natural antioxidative compounds. It can be concluded that all of the extracts can be used as an accessible source of natural antioxidants with consequent health benefits. Nevertheless, it is suggested to conduct further studies to ensure the medicinal properties of the gastropods along with other bioactivities such as anti-inflammatory, cytotoxicity, anticancer, antimalarial, analgesic activity, antiallergic and antihypertensive activity.

REFERENCES

Anand, P.T., Chellaram, C., Kumaran, R. and Shanthini, C. F. (2010). Biochemical composition and antioxidant activity of *Pleuroploca trapezium meat*. J. Chem. Pharm. Res., 2: 526-535.

Brand-Williams, W., Cuvelier, M. E., and Berset, C. (1995) Use of a free radical method to evaluate antioxidant activity. *LWT*—Food Science and Technology. 28(1): 25–30.

Benkendorff, K., C.M. McIver and C.A. Abbott (2011). Bioactivity of the murex homeopathic remedy and of extracts from an Australian murcid molluks against human cancer cells. Evidence-Based Complementary and Alternative Medicine, Article ID 879585, 12 pages. https://doi.org/10.1093/ecam/nep042

Bhakuni, D. S. and Rawat, D. S. (2005). Bioactive Marine Natural Products. Springer, New York and Anamaya Publishers, New Delhi, India. p 26-63.

Bouchet, P. & J.-P. Rocroi (2005). Classification and nomenclator of gastropod families. Malacologia 47: 1–397.

Chellaram, C. and Edward. J. K. P. (2009). Antinociceptive assets of coral associated Gastropod, *Drupa margariticola*. Int. J. Pharmacol., 5: 236-239.

Defer, D., N. Bourgnon and Y. Fleury (2009). Screening for antibacterial and antiviral activities in three bivalve and two gastropod marine mollusks. Aquaculture. 293: 1-7.

Deny Susanti, Hasnah M. Sirat, Farediah Ahmad, Rasadah Mat Ali, Norio Aimi, Mariko Kitajima (2007). Antioxidant and cytotoxic flavonoids from the flowers of Melastoma malabathricum L. Food Chem. 107(3) 710-716

Houssen, W. E. and Jaspars, M. (2005). Natural Products Isolation, Second Edition, Methods in Biotechnology, Humana Press, 20, 353-390.

Huang, D. J., Chen, H. J., Lin, C. D. &Lin, Y. H. (2005). Antioxidant and antiproliferative activities of water spinach (Ipomoea aquatic Forsk) constituents. *Bot. Bull. Acad. Sin.* 46, 99-106.

Malve, H (2016). Exploring the ocean for new drug developments: marine pharmacology. J. Pharm. Bioallied Sci. 8(2): 83-91. Doi: 10.4103/0975-7406.171700

Molyneux, P. (2004). The use of the stable free radical diphenylpicrylhydrazyl (DPPH) for estimating antioxidant activity. Songklanakarin. *J. Sci. Technol.* 26, 211-219.

Xiu, R. Y., Yi, . Q., Yu, Q. Z., Chang, F. C. and Wang, B. (2019). Purification and characterisation of antioxidant peptide derived from protein hydrolysate of marine bivalve mollusk Tergillarca granosa. Mars Drugs. 17(5), 251-266.

Nagash, Y.S., R.A Nazeer, and N.S. Sampath Kumar (2010). In vitro antioxidant activity of solvent extracts of mollusks (Loligo duvauceli and Donax strateus) from India. World J. Fish. Mar. Sci., 2: 240-245.

Rahman, M. M., Islam, M. B., Biswas, M. and Alam, A. H. M. K. (2015). In vitro antioxidant and free radical scavenging activity of different parts of Tabebuia pallida growing in Bangladesh. BMC Res. Notes. 8: 621. DOI 10.1186/s13104-015-1618-6

Siraprapha, P., Soranan, W. and Pobporn, T. (2016). Molluscan Fauna in Bang Taboon Mangrove Estuary, Inner Gulf of Thailand: Implications for conservation and sustainable use of coastal resources: p. 1-5. MATEC Web of Conferences. CCBS 2016.

Sies H (1997). Oxidative stress: oxidants and antioxidants. *Exp Physiol* 82 (2): 291–295.

Siong Kiat Tan and Reuben Clements (2008) Taxonomy and distribution of the Neritidae (Mollusca: Gastropoda) in Singapore. Zoological Studies 47(4): 481-494.

Soler-Rivas, C., Espin, J.C. and H.J. Wichers (2000). An easy and fast test to compare total free radical scavenger capacity of foodstuffs. *Phytochem. Anal.* 11, 330-338.

Solé, M., Porte, C., Albaigés, J. (1994) Mixed function oxygenase system components and antioxidant enzymes in different marine bivalves: its relation with contaminant body burdens. Aquat Toxicol 30:271–283

Tan, S. K. and Clements, R. (2008) Taxonomy and distribution of the neritidae (Mollusca: Gastropoda) in Singapore.

Tinu, Odeleye, William Lindsey and White, Jun Lu (2019). Extraction techniques and potential health benefits of bioactive compounds from marine mollusks: a review. Journal of Food Function. 22:10(5):2278-2289.

Subhapradha, N., Ramasamy, P., Sudharsan, S., Seedevi, P., Moovendhan, M., Dharmadurai, D., Vasanth Kumar, S., Vairamani, S. and Shanmugam, A. (2013) Antioxidant potential of crude methanolic extract from whole body tissue of *Bursa spinosa*. Proceedings of the National conference-USSE-2013, TBML College, Porayar-609307, Nagai-Dt, Tamil Nadu, South India. 163-167.

Heavy metal resistant bacteria from marine sediment of Pantai Balok, Pahang, Malaysia

Munira Haniff[1], Zaima Azira Zainal Abidin[1]*

[1]*Dept. of Biotechnology, Kulliyyah of Science, International Islamic University Malaysia*
Corresponding author: zzaima@iium.edu.my

ABSTRACT

Heavy metal pollution particularly in the coastal water has become an issue of serious international concern. Heavy metal pollution not only affect the quality of water and soil, but it also affects the animals and plants as well as the microorganisms inhabiting the coastal area. This study aimed at isolation of heavy metal resistance bacteria from marine sediment of Pantai Balok as an attempt to assess possible heavy metal pollution present in that area as well as in search of potential candidates for bioremediation purposes. A total of 33 isolates were obtained and subjected for heavy metal resistance test using the following heavy metals - chromium (Cr), nickel (Ni), copper (Cu), cobalt (Co), cadmium (Cd). Results revealed that almost all isolates showed high tolerance towards Cr, Ni, Co and Cu but low tolerance towards Cd. Heavy metal resistance profile associated with Pantai Balok was in the following order: Cr > Ni > Co > Cu > Cd. Five isolates namely PB1, PB9, PB17, PB18, and PB 33 exhibited strong heavy metal resistance pattern and their identities were determined using 16S rRNA gene sequencing. PB1 was closely related to *Stenotrophomonas maltophilia* (99%) while PB9 to *Staphylococcus pasteuri* (98%). Isolates PB17 and PB18 were highly similar to *Bacillus pumilus* (99%) and *Bacillus sp.* (99%) respectively whereas PB33 is *Pseudomonas aeruginosa* (99%). The presence of heavy metal resistant bacteria may indicate the occurrence of heavy metal pollution in Pahang coastal water and may pose a potential health risk to the public.

Keywords: Heavy metal resistant, Bacteria, Marine sediment, 16S rRNA gene

INTRODUCTION

Expanding in urbanization activities nowadays has made the coastal area become an unhealthy condition where numerous chemicals such as heavy metals and pesticides have been used and discharged to the coastal area. Heavy metals are one of the major sources of environmental pollution due to its discharge effluent into the environment by a huge number of industrial activities such as metal processing, mining and others (Yang et al. 2018; Yamina et al. 2012). Heavy metal is any metal or metalloid of environmental concern that also has toxic chemical elements and their deviated chemical compounds. It has density criteria ranging from above 3.5 g/cm3 to above 7 g/cm3 (Nies 1999). Nevertheless, it is still undeniable that some of these heavy metals are necessary for life, such as copper, iron and zinc. However, other heavy metals such as arsenic, cadmium, mercury and silver have no biological role in organisms, and they are harmful even at very low concentrations (Alam et al. 2011). In an aquatic environment, heavy metals tend to accumulate in the sediment. As heavy metals are rapidly discharged into the environment, it will associate with particulates and finally settles at the bottom of the sediments (Chapman et al. 1998). Furthermore, heavy metal pollution in the marine environment is becoming a concern due to its ability to accumulate in the food chain. Moreover, many human activities have resulted in the accretion of metals in the environment and finally are accumulated through the food chain and lead to serious health and ecological problems (Mohammadi et al. 2019; Vareda et al. 2019; Hou et al. 2018; Deng and Wang 2012).

Microorganisms are very sensitive to low concentration of heavy metals, however, due to certain specific habitat conditions they can rapidly try to adapt those changes and become resistant to high content of heavy metals (Nithya and Pandian, 2009). Microorganisms respond to heavy metals by various operations; including transport across the cell membrane, biosorption to the cell walls and entrapment in extracellular capsules, precipitation, complexation, oxidation-reduction reactions, production of extracellular cultures, intracellular sequestration, metal efflux pumps and biomineralization (Álvarez et al. 2013; Schütze and Kothe 2012). The ability of microorganisms to survive and reproduce in a metal-contaminated habitat depends on the adaptation of genetic or physiological as the heavy metal resistance bacteria is commonly encoded by genes or plasmids and transposons and they can regularly transferable intergenerically, interspecifically from in situ microflora to indigenous microflora (Malik and Aleem 2011). Examples of heavy metal resistance genes (MRGs) include copper resistance genes (*cop*A, *cop*B, *pco*A, *pco*C, and *pco*D), arsenic resistance genes (*ars*B and *ars*C), nickel, lead and chromium resistance gene *(ncc*A, *pbr*T, and *chr*B respectively) (Chen et al. 2019).

Pantai Balok is a famous beach that is located at the South China Sea and is considered as one of the attractions for tourists in Pahang alongside Teluk Chempedak and Pantai Batu Hitam. However, it is highlighted that the coastal area was being polluted with dumping of waste and badly monitored. Anthropogenic activities such as land use for development in the coastal zone, untreated domestic and industrial waste effluent, oil spill incidents or unlawful discharge oil effluent may contribute to the marine pollution of Pahang state coastline. The status of heavy metal resistant bacteria in sediment of Pantai Balok is relatively unknown as no study was conducted in this area. Hence, this study provides an insight of heavy metal resistant bacteria present in the marine sediment of Pantai Balok. Moreover, identification of heavy metal resistance bacteria can be utilised as biological indicators of heavy metal contamination and candidates for bioremediation application in the future.

MATERIALS AND METHODS
Sediment sample collection
Marine sediment samples were collected using a Ponar grab from the Balok beach area at three different stations namely Station 1, Station 2 and Station 3. Table 1 describes the coordinates, depth and pH of the sampling area. Each of the stations was located 30 m apart from one another. All collected sediment samples were transferred to a sterilized polyethylene plastic bag and processed immediately.

Table 2.1: Sampling station and coordinates of Pantai Balok area

Location	Coordinates	Depth	pH
Station 1	N 03 '55.768 E 103' 23.395	4.2 m	6.9
Station 2	N 03'56.115 E 103'23.536	3.4 m	6.0
Station 3	N 03'56.397 E 103'23. 660	3.4 m	6.6

Isolation of bacteria from marine sediment samples
Bacteria from sediment samples were isolated using the spread plate technique (Zainal Abidin et al. 2018). One gram of sediment samples was mixed with 10 ml of saline solution. Then, the homogenized samples were serial diluted (10^{-2} to 10^{-5}) and 100 µl of each dilution was plated on the nutrient agar in duplicate. The plated samples were then incubated for 48 hours at 37°C. After incubation, respective colonies were purified on nutrient agar medium. Gram staining was performed on all isolates and their morphological characteristics were recorded.

Heavy metal resistance test
The heavy metal resistance of the bacterial strains obtained was determined using Mueller Hinton agar supplemented with various concentrations of five different heavy metals (Cd^{2+}, Cu^{2+}, Cd^{2+}, Co^{2+}, Ni^{2+}) in the form of chloride salts. The initial concentration of the heavy metal was at 20 µg/ ml and the concentration of the heavy metals was gradually increased at 10 µg/ml until the isolates failed to grow. Minimum Inhibitory Concentration (MIC) was noted when the isolates failed to grow on plates even after a maximum of 5 days of incubation. The test was conducted in duplicate.

Polymerase chain reaction (PCR) amplification 16S rRNA gene
Isolates displaying heavy metal resistance capability were subjected to molecular identification using 16S rRNA gene sequence. Genomic DNA of isolates was extracted using GF-1 bacterial DNA Extraction Kit (Vivantis) following manufacturer's protocols. PCR amplification of the 16S rRNA gene was conducted using the following set of primers: 27F 5'–AGAGTTTGATCCTGGCTCTCAG–3' and 1492R 5'–GGTTACCTTGTTACGACTT–3'. The PCR reactions were performed in a final volume of 50 μl which consist of 200 ng DNA template, 25 μl of MyTaq™ Mix 2X (Bioline, UK) and 0.4 μM primers under the following conditions: initial denaturation at 94°C for 5 min, followed by 30 cycles of 94°C for 30 s, 55°C for 60 s and 72°C for 4 min; and extension step at 72°C for 10 min. The amplification products were confirmed using 1% agarose gel and sent to 1st Base Laboratory, Malaysia for purification and sequencing. The resultant 16S rRNA gene sequences were manually verified and edited using the BioEdit sequence alignment editor. The partial nucleotide sequences analysis of the isolates was carried out via the GenBank BLASTn search tool.

RESULTS AND DISCUSSION
In total, 33 isolates were obtained from 3 sampling points and majority (~75%) of the isolates belonged to Gram negative bacteria (Table 2). Most bacterial colonies were found to be white and cream in color with few isolates displaying other colors such as peach, yellow and orange. Colony morphology and Gram staining of representative isolates from each sampling point are as illustrated in Figures 1-3.

Table 2.2: Distribution of Gram positive and Gram-negative bacteria according to the sampling points

Location	Gram positive bacteria	Gram negative bacteria
Point 1	10	3
Point 2	9	3
Point 3	6	2
Total	**25**	**8**

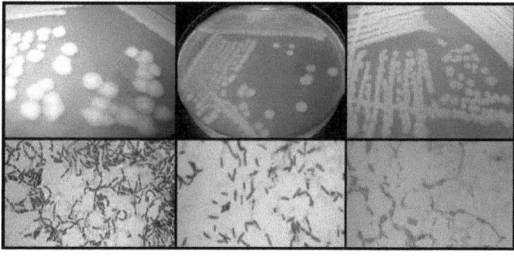

Fig. 1: Representative isolates from Point 1

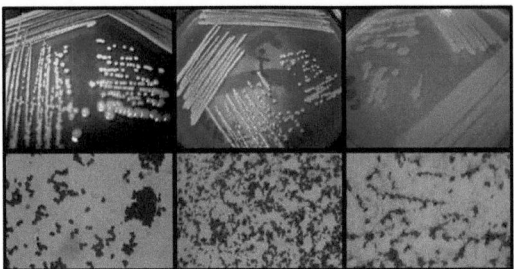

Fig. 2: Representative isolates from Point 2

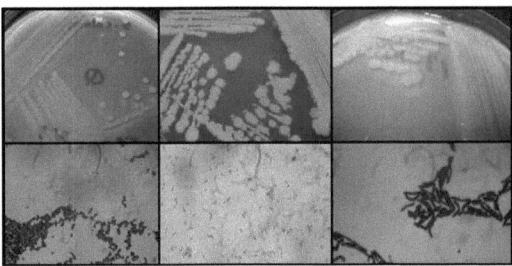

Fig. 3 Representative isolates from Point 3

In this study, all isolates were found to have MIC > 450 µg/ml for Cr indicating that these bacteria possessed strong tolerance towards Cr (Table 3). A few studies demonstrated that some of the bacteria isolated may possibly tolerate a concentration of Cr up to 1,000 µg/ml (Sair and Khan 2017; Yamina et al. 2012). As for Ni, almost all isolates showed MIC > 450 µg/ml except for PB5 and PB24, whereby MIC for both isolates were 450 µg/ml. Two third of the isolates showed MIC > 450 µg/ml for Co while the MICs for the rest of the isolates were in the range of 200 – 400 µg/ml. Majority of the isolates (72.7%) recorded MIC > 450 µg/ml for Cu while the remaining were in the range of 100 – 400 µg/ml. Heavy metal resistant bacteria are considered as the biological indicators of heavy metal contamination of a particular location. Moreover, such bacteria potentially contribute to biogeochemical cycling of heavy metal in the environment. High tolerance to Cr, Ni, Cu and Co by majority of the bacterial isolates may suggest the possibility of contamination by these heavy metals occurred in Pantai Balok. The fact that Gebeng industrial area just miles away from Pantai Balok may also be a contributing factor to this observation. Unusual resistance to Cr may relate to contamination by Cr in that particular area. Cr is widely used in industry as plating, alloying, tanning of animal hides, textile dyes and mordants and these activities consequently led to increased environmental contamination of Cr (Oliveira 2012).The presence of Ni in the marine sediment of Pantai Balok may be linked to industrial effluent, land application of fertilizers, wastewater irrigation and sewage sludge. High concentration of Ni will lead to high number of nickel-resistant strains in the bacterial community inhabiting the marine sediment (Mengoni et al. 2001). Co pollution may be attributed to emission of cobalt compounds during combustion of hard coal and petroleum, petrochemical, metals and ceramic industries resulted in substantial Co accumulation in marine sediment (Kosiorek and Wyszkowski 2019). Cu contamination usually occurred because of the application of agriculture input due to the fact that Cu is an essential micronutrient important for the growth of plants specifically in disease resistance and production of seed (Wuana and Okiemen 2011). Results obtained from this study also indicated multiple heavy metal resistant capability of these bacteria. Bacteria resistant to particular heavy metal might also acquire resistance to other heavy metals. Previously, Cr (VI) resistant bacteria from the high Cr-contaminated sites were also showing resistance to Cr(III), Ni, Zn, Cu, Cd and Hg (Alam et al., 2011; Verma et al., 2001). Similarly, investigation on the abundance of metal resistance genes (MRGs) in a copper tailing dam area found the presence of multiple heavy MRGs that are encoded by *czc*A, *czc*C, and *czc*D (Chen et al. 2019). There are two different mechanisms of co-selection regulate for the multiple heavy metal resistance which include co-resistance where genetically linked different resistance factors transfer concurrently and cross-resistance in which the same factor is responsible for resistance to more than one structurally dissimilar compound (Baker-Austin et al., 2006).

Although almost all isolates showed high tolerance towards Cr, Ni, Co and Cu, these isolates however, displayed low tolerance towards Cd. The highest MICs recorded for Cd were 300 µg/ml and 280 µg/ml by isolates PB17 and PB33. The lowest MIC recorded was 70 µg/ml from 3 isolates (PB7, PB8, P23) while MICs for the remaining of the isolates were in the range 100 -140 µg/ml. Similar observation was also reported by Zainal Abidin and Chowdhury (2018) in Teluk Chempedak and Pantai Batu Hitam, both of which are located on the Pahang coastline. Cd is widely applied in many industries like paints, electroplating, and copper alloys, pulp and paper, alkaline batteries, and mining, fertilizer and zinc refining (USEPA 2000). Since all isolates displayed low tolerance to Cd, this observation may indicate that Pantai Balok is not polluted with Cd. The resistance pattern associated with Pantai Balok, Pahang was in the form of Cr > Ni > Co > Cu > Cd.

Table 3: MIC of heavy metal in µg/ml

Isolate	Chromium (µg/ml)	Cobalt (µg/ml)	Copper (µg/ml)	Cadmium (µg/ml)	Nickel (µg/ml)
PB1	>450	>450	>450	140	>450
PB2	>450	250	400	120	>450
PB3	>450	250	>450	120	>450
PB4	>450	400	>450	140	>450
PB5	>450	300	>450	130	450
PB6	>450	>450	250	100	>450
PB7	>450	>450	>450	70	>450
PB8	>450	>450	>450	70	>450
PB9	>450	>450	>450	140	>450
PB10	>450	250	>450	100	>450
PB11	>450	250	>450	120	>450
PB12	>450	450	>450	120	>450
PB13	>450	400	>450	100	>450
PB14	>450	>450	100	100	>450
PB15	>450	>450	>450	100	>450
PB16	>450	>450	>450	120	>450
PB17	>450	>450	>450	300	>450
PB18	>450	>450	>450	140	>450
PB19	>450	>450	>450	100	>450
PB20	>450	>450	>450	100	>450
PB21	>450	>450	>450	120	>450
PB22	>450	400	>450	140	>450
PB23	>450	200	150	70	>450
PB24	>450	200	150	100	450

PB25	>450	>450	250	100	>450
PB26	>450	>450	200	100	>450
PB27	>450	>450	>450	100	>450
PB28	>450	>450	>450	100	>450
PB29	>450	>450	300	100	>450
PB30	>450	>450	>450	100	>450
PB31	>450	>450	250	100	>450
PB32	>450	>450	>450	100	>450
PB33	>450	>450	>450	280	>450

Table 4: Identities of isolates exhibiting high resistance to heavy metals

Isolate	MIC of heavy metal in µg/ml					Closest relative	Similarity (%)
	Cr^{2+}	Co^{2+}	Cu^{2+}	Cd^{2+}	Ni^{2+}		
PB1	>450	>450	>450	140	>450	*Stenotrophomonas maltophilia* strain SJTH1	99%
PB9	>450	>450	>450	140	>450	*Staphylococcus pasteuri* strain AE4-2	98%
PB17	>450	>450	>450	300	>450	*Bacillus pumilus* strain NCTC10337	99%
PB18	>450	>450	>450	140	>450	*Bacillus* sp. strain C81	99%
PB33	>450	>450	>450	280	>450	*Pseudomonas aeruginosa* strain C-1	99%

Molecular identification through PCR amplification of 16S rRNA gene was performed on 5 isolates – PB1, PB9, PB17, PB18 and PB33, all of which displayed strong heavy metal resistance profiles. All 5 isolates gave high readings (>450 µg/ml) for Cr, Ni, Co and Cu and 3 isolates (PB1, PB9 and PB18) gave quite low MIC for Cd (140 µg/ml) whereas PB33 and PB17 had MIC values for Cd of 300 µg/ml and 280 µg/ml respectively. PCR amplification of the 16S rRNA gene (~1,500 bp) for these isolates was successfully obtained and the partial sequences of the 16S rRNA gene were compared with the NCBI database (Table 4). Partial sequence of 16S rRNA gene indicated that PB1 was closely related to *Stenotrophomonas maltophilia* with 99% similarity while PB9 is highly similar to *Staphylococcus pasteuri* (Table 4). *Stenotrophomonas maltophilia* is a Gram-negative bacterium and *S. maltophilia* strains are found to be ubiquitously distributed in the environment including coastal waters. *S. maltophilia* may cause nosocomial infections in immune-compromised patients and are naturally resistant to many broad-spectrum antibiotics such as cephalosporins, carbapenems, and aminoglycosides. Several studies have reported the incidence of heavy metal resistant *S. maltophilia* (Baldiris et al. 2018; Raman et al. 2018; Pages et al. 2008), indicating the heavy metal resistant capability of this bacterium alongside antibiotic resistance. Isolates PB17 and PB18 were found to belong to genus *Bacillus* with PB17 closely related to *B. pumilus* whereas isolate PB33 is identified as *Pseudomonas aeruginosa* based on partial sequence of 16S rRNA gene. This finding is in accord with other findings (Zainal Abidin et al. 2020; Dweba et al. 2019; Pereira and Ramaiah 2019; Verma et al. 2017; Fierros-Romero et al. 2016) that demonstrated *Bacillus* sp., *Pseudomonas* sp. and *Staphylococcus* sp. strains possess the multiple-metal resistance ability. *Bacillus* sp. is a Gram positive, rod-shaped bacterium and can be isolated from various environments including human and animals. Jayanthi et al. (2016) reported the occurrence of *B. pumilus* to be absolutely resistant to a range of heavy metals (Pb, Hg, Cd, Cr. Mn, Zn, Al, Fe). *P. aeruginosa* is a Gram-negative bacterium widely distributed in the environment, as well as in various living host organisms. Besides, this bacterium is the most prevalent cause of opportunistic infections in humans. *P. aeruginosa* commonly shows resistance to multiple antibiotics and this bacterium is also known to have heavy metal resistance ability. For instance, *P. aeruginosa* ASU 6a isolated from a highly metal polluted habitat was found to exhibit high degree of tolerance to Pb^{2+}, Cd^{2+}, Cr^{6+} and Ni^{2+} and resistant to several antibiotics (Hassan et al. 2008). *S. aureus* is one of the most important pathogens of humans and animals. MRSA (Methicillin resistant *S. aureus*) is a notorious pathogen, a common cause in hospital infection and is resistant to multiple antibiotics. Research conducted by Dweba et al. (2019) found that *S. aureus* isolates were found to be resistant to high concentration of Cd, Zn, Pb and Cu. All 5 isolates have the potential to be utilized in biotechnological application and further investigation is required to fully capitalize their capabilities as biological test tools in heavy metal contaminated sites and in application in bioremediation of heavy metal polluted areas.

CONCLUSION
The presence of bacteria with high tolerance to Cr, Ni, Co and Cu from marine sediment of Pantai Balok may suggest the existence of heavy metals contamination at this site. These findings exemplify the impact of human activities on marine environments which can constitute a risk to the public health and pose a threat to the marine ecosystem. Relevant parties including local communities may need to impose monitoring and enforcement in the Pahang coastline with the aim to reduce the impact of anthropogenic activities on marine ecosystems. Additionally, five isolates (PB1, PB9, PB17, PB18 and PB33), all of which exhibited strong heavy metal resistance have the potential to be employed in biotechnological application particularly in bioremediation of heavy metal polluted sites.

REFERENCES
Alam M.Z., Ahmad S., Malik, A. (2011). Prevalence of heavy metal resistance in bacteria isolated from tannery effluents and affected soil, *Environ. Monit. Assess.,* 178: 281–291.

Álvarez, A., Catalano, S.A., Amorosono, M.J. (2013). Heavy metal resistant strains are widespread along *Streptomyces* Phylogeny. *Molecular Phylogenetics and   Evolution,* 66:1083-1088.

Baker-Austin, C., Wright, M. S., Stepanauskas, R., J.V.McArthur, J.V. (2006). Co-selection of antibiotic and metal resistance. *Trends in Microbiology, 14(4):* 176-182.

Baldiris, R., Acosta-Tapia, N., Montes, A., Hernández, J., Vivas-Reyes, R. (2018). Reduction of Hexavalent Chromium and Detection of Chromate Reductase (ChrR) in *Stenotrophomonas maltophilia*. *Molecules,* 23:408 doi:10.3390/molecules23020406

Chapman, P.M. Wang, F. Janssen, C. Persoone G., Allen, H.E. (1998). Ecotoxicology of metals in aquatic sediments binding and release, bioavailability, risk assessment, and remediation. *Can. J. Fish. Aquat Sci.,* 55: 2221-2243.

Chen, J., Li, J., Zhang, H., Shi, W., Liu, Y. (2019). Bacterial Heavy-Metal and Antibiotic Resistance Genes in a Copper Tailing Dam Area in Northern China. *Front. Microbiol.* 10:1916. doi: 10.3389/fmicb.2019.01916

Deng, X., Wang, P. (2012). Isolation of marine bacteria highly resistant to mercury and their bioaccumulation process. *Bioresource Technology*, 121: 342–347.

Dweba, C.C., Zishiri, O.T., El Zowalaty, M.E. 8 (2019) Isolation and molecular identification of virulence, antimicrobial and heavy metal resistance genes in Livestock-associated methicillin resistant *Staphylococcus aureus*, *Pathogens*, 1–21.

Fierros-Romero, G., Gómez-Ramírez, M., Arenas-Isaac, G.E., Pless, R.C., Rojas-Avelizapa, N.G. (2016). Identification of *Bacillus megaterium* and *Microbacterium liquefaciens* genes involved in metal resistance and metal removal, *Can. J. Microbiol.,* 62: 505–513.

Hassan, S., H. A., Abskharon R. N. N., Gad El-Rab, S. M. F., Shoreit A. A. M. (2008). Isolation, characterization of heavy metal resistant strain of Pseudomonas aeruginosa isolated from polluted sites in Assiut city, Egypt, *Journal of Basic Microbiology*, 48:168–176.

Hou, S., Zheng, N., Tang, L., Ji, X., Li, Y., Hua, X. (2018). Pollution characteristics, sources, and health risk assessment of human exposure to Cu, Zn, Cd and Pb pollution in urban street dust across China between 2009 and 2018. *Environment International*, 128, 430-437.

Jayanthi, B., Emenike C.U., Agamuthu, P., Khanom Simarani, Sharifah Mohamad, Fauziah, S.H. (2016). Selected microbial diversity of contaminated landfill soil of Peninsular Malaysia and the behavior towards heavy metal exposure, *Catena* 147: 25-31

Malik, A., Aleem, A. (2011). Incidence of metal and antibiotic resistance in *Pseudomonas* spp. from the river water, agricultural soil irrigated with wastewater and groundwater. *Environ Monit Assess* 178: 293–308.

Mengoni, A. Barzanti, R. Gonnelli, C. Gabbrielli, R. Bazzicalupo, M. (2001). Characterization of nickel-resistant bacteria isolated from serpentine soil, *Environ. Microbiol.*, 3, 691–698.

Mohammadi, A.A., Zarei, A., Esmaeilzadeh, M., Taghavi, M., Yuosefi, M., Yousefi, Z., Sedighi, F., Javan, S. (2020) .Assessment of Heavy Metal Pollution and Human Health Risks Assessment in Soils Around an Industrial Zone in Neyshabur, Iran, *Biol Trace Elem Res,* 195, 343–352

Nies, D.H., 1999. Microbial heavy-metal resistance. *Appl. Microbiol. Biotechnol.* 51, 730–750.

Nithya, C., Pandian, S. K. (2010). Isolation of heterotrophic bacteria from Palk Bay sediments showing heavy metal tolerance and antibiotic production. *Microbiological Research*, 165(7), 578–593.

Pages, D., Rose, J., Conrod, S., Cuine, S., Carrier, P. (2008) Heavy Metal Tolerance in *Stenotrophomonas maltophilia*. *PLoS ONE* 3(2): e1539. doi:10.1371/journal.pone.0001539

Pereira, E.J., Ramaiah N. (2019). Chromate detoxification potential of *Staphylococcus* sp., Isolates from an estuary, *Ecotoxicol.*, 28: 457–466.

Raman, N., Asokan, M., Shobana, S. Sundari, N. (2018). Bioremediation of chromium (VI) by *Stenotrophomonas maltophilia* isolated from tannery effluent. *Int. J. Environ. Sci. Technol.* 15: 207–216

Sair A.T., Khan, Z.A. (2017) Prevalence of antibiotic and heavy metal resistance in gram-negative bacteria isolated from rivers in northern Pakistan, *Water Environ. J.*, 32: 51–57.

Schütze, E., Kothe, E. (2012). Bio-Geo Interactions in Metal-Contaminated Soils. In: Kothe, E., Varma, A. (Eds.), *Soil Biology* 31. Springer-Verlag, Berlin Heidelberg, pp. 163–182.

USEPA (2000) Introduction to phytoremediation. United States Environmental Protection Agency, Washington.

Vareda, J.P., Valente, A.J.M., Durães, L. (2019). Assessment of heavy metal pollution from anthropogenic activities and remediation strategies: A review. *Journal of Environmental Management,* 246, 101–118.

Verma, G., Christy, N., Veer, C. (2017). Isolation and Characterization of Pseudomonas stutzeri as lead tolerant Bacteria from water bodies of Udaipur, India using 16S rDNA sequencing technique, J. *Pure Appl. Microbiol.*, 11: 975–979

Wuana, R. A., Okieimen, F. E. (2011). Heavy metals in contaminated soils: a review of sources, chemistry, risks and best available strategies for remediation. *ISRN Ecology*, *2011*.

Yamina, B., Tahar, B., & Laure, F. M. (2012). Isolation and screening of heavy metal resistant bacteria from wastewater: A study of heavy metal co-resistance and antibiotics resistance. *Water Science and Technology: A Journal of the International Association on Water Pollution Research*, 66(10), 2041–8.

Yang, Q., Li, Z., Lu, X., Duan, Q., Huang, L., Bi, J. (2018). A review of soil heavy metal pollution from industrial and agricultural regions in China: Pollution and risk assessment. *Science of The Total Environment,* 642, 690-700.

Zainal Abidin, Z.A., Chowdhury, A.J.K. (2018). Heavy Metals and Antibiotic Resistance Bacteria In Marine Sediment Of Pahang Coastal Water. *J. CleanWAS*, 2(1): 20-22.

Zainal Abidin, Z.A., Badaruddin, P.N.E., Chowdhury, A.J.K. (2020) Isolation of heavy metal resistance bacteria from lake sediment of IIUM, *Kuantan Desalination and Water Treatment* 188: 431–435.

SALINITY TOLERANCE AND GROWTH PERFORMANCE OF ASIAN SEABASS (*Lates calcarifer*) JUVENILES

Kim Seng, Tan[1], Mohammad Tajuddin Abd Manaf[1], Najiah Musa[1], Kok Leong, Lee[1], Nadirah Musa[1*]

[1]*Faculty of Fisheries and Food Science, Universiti Malaysia Terengganu, 21030 Kuala Nerus, Terengganu*

corresponding author: nadirah@umt.edu.my

ABSTRACT

The current study aims to determine the tolerance and growth rates of Asian seabass juveniles subjected to different range of water salinities i.e. 0, 5, 10, 15, 20, 25 and 30ppt. The fish were also subjected to growth performance study for 15 days. No mortality was observed during the experimental period. A significantly higher growth performance of total length gain (TLG), total weight gain (TWG) and specific growth rate (SGR) were observed at 0 and 25 ppt with 6.16 and 8.08%; 29.94 and 26.92% and 1.72 and 1.58%, respectively. Overall, Asian seabass juveniles reared at 0 ppt salinity for 15 days attained a better value for TWG and SGR when compared to 25 ppt. Therefore, manipulating the salinity levels may be beneficial for hatchery management in order to increase the survival and production of Asian seabass.

Keywords: *Lates calcarifer*; salinity tolerance; growth performance

INTRODUCTION

During the last few decades, the fishery sector holds a large potential to provide an important source of protein to the Malaysian population. According to FAO (2018), the total fishery production of the country amounted to 1.7 million tonnes with the total value of exports earnings of USD714.1 million in 2017. In general, the fisheries can be divided into two major components, i) marine capture fisheries, and; ii) aquaculture. However, capture fisheries is the highest contributing sector of the fish landings which created 88.3 % of the total production in 2007, while the rest coming from aquaculture (FAO, 2018).

Asian Seabass, *Lates calcarifer* locally known as "ikan siakap" is a tropical and sub-tropical member of the family Latidae of order Perciformes (Shadrin and Pavlov, 2015).This fish is widely distributed throughout the Indo-West Pacific region from Arabian Gulf to southern China, Papua New Guinea and northern Australia (Nelson, 1994). The price of Asian Seabass at the local market has risen to as much as RM16 per kilogram. The demand of Asian Seabass is considered high and is one of the most popular fish among Malaysians due to its fine texture and tasty whiteflesh.

Lates calcarifer spawns all the year round in nature, with the peak season occurring from April to August. The adult fish is a voracious carnivore, but juveniles are omnivorous (Kungvankil et al., 1985). They seem to require saline water during the spawning season, yet the larvae can also be found in freshwater. The larvae metamorphosed to fry at 8-10 mm which can be recognized easily by the change in color of the fish larvae from dark into brownish and the appearance of distinct lateral stripes (Dhert, Laven & Sorgeloos, 1992); and later change to fingerling stage at 2 to 3 weeks of age (20 mm).

Aquaculture especially of brackish water fish culture in Malaysia has potential for development. As such, Asian seabass is an important coastal, estuarine and freshwater fish has been the target of culture species for local fish farmers because of its high market value and fast growth rate (FAO, 2018). Nevertheless, the success of seed production starts from the availability of healthy broodstock and the consistency of high quality of mass seed production. However, at present, the quality of the seed of Asian Seabass is inconsistent while inadequate supply of seed has been reported either from the wild or the aquaculture (Nammalwar and Marichamy, 1998).

Various physiology processes in fish such as metabolism, osmoregulation and biorhythm are affected by water salinity. Besides that, salinity affects distribution, growth and survival rate of the fish development (Varsamos et al., 2005). Bony fish can maintain environmental salinities of their body fluids in the ionic and osmotic homeostasis through energy demanding processes of the osmoregulatory mechanisms (Sampaio and Bianchini, 2002). Growth is the net positive result from the energy provided by food ingestion and the metabolic expenditure (Jobbling, 1994). It has been reported that when the salinity is at the optimum level, the net energy can help to enhance the fish growth rates (Amni   et al., 2015) and reduce the osmotic work (Estudillo et al., 2000). Yet, only a few studies have been performed to investigate the salinity tolerance of Asian Seabass. Therefore, this experiment was conducted to determine the salinity tolerance and growth rates of Asian Seabass (*Lates calcarifer*) juvenile subjected to various salinity treatment.

MATERIALS AND METHODS
Source of juveniles Asian Seabass
Asian Seabass, *Lates calcarifer* juveniles (50 days post hatching) were bought from a local supplier. Each of the juveniles was measured for body weight and body length (mean body weight 11.80 ± 3.75g; mean body length of 10.26 ± 1.15cm). Experiments were performed at the marine hatchery, Hatchery Unit, Faculty of Fisheries and Food Science, Universiti Malaysia Terengganu.

Experimental setup
Sea water was stored in the reservoirs and filtered through biological filters or rapid sand filters to maintain the required water quality. Different salinity waters were prepared (5, 10, 15, 20 (control), 25 and 30 ppt) and diluted with freshwater and kept in a closed glass aquarium. Freshwater was used for 0 ppt. Fourteen units of 54 litre volume of glass aquarium (60 cm × 30 cm × 30 cm depth) were prepared and washed before the start of the experiment and filled with water of different levels of salinity. Refractometer was used to measure the salinity of the water used. In addition, a gentle aeration was placed into the aquarium to improve the water circulation; and to provide with continuous dissolved oxygen.

One hundred and forty healthy juveniles of seabass of similar sizes were transferred to a stocking tank (210 cm × 120 cm × 74 cm depth) of 350 L, filled with aerated water of 20 ppt for acclimation of 1 week. On arrival, the fish were initially starved and subjected to 10ml of 5 ppm iodine for initial treatment in 5 hours, and continued for acclimation with 20 ppt. After 24h, the fish were fed twice daily with commercial marine fish pellets (43% crude protein, 6% crude fat and 12% moisture) at 2.0% b.w.

Salinity Tolerance
The first experiment was conducted to determine the effect of water salinity on the survival rate of juvenile Asian Seabass. Prior to the salinity trials, the juveniles were starved for 24h. Their total length (TL) and body weight (BW) were recorded. Glass aquariums of different water salinity were prepared; in replicates. A total of 70 juveniles were equally distributed into 14 aquariums (n=5) and kept for 48 hours. The fish were not fed during the trials, mortality was observed daily, and dead fish was removed.

Effect of salinity on growth performance
No mortality was recorded during the salinity tolerance trials. Therefore, water salinities of 0, 5, 10, 15, 20 (control), 25 and 30 ppt were used for the growth performance experiment that lasted for 15 days. The experiments were conducted in replicates (Amornsakun *et al.*, 2016). The total length (TL) and total weight (TW) of 70 fish were measured and recorded prior to the experiment and at the end of the 15 days period. Seventy seabass juveniles were equally distributed into 14 aquariums (n=5).

Water quality parameters such as temperature, salinity, dissolved oxygen, pH and mortality were recorded daily. The immersion heaters were used to maintain the water temperatures at $28 \pm 1°C$. Each of the aquariums was provided with aeration to maintain the dissolved oxygen saturation levels in the range of

60-70%. During the experiment the juveniles were fed twice daily at 2% of body weight with commercial marine fish pellets. Feces and uneaten feed waste were siphoned out of the aquariums daily. During the 15 days period, one third of the water volume was replaced every 3 days just before the feeding time.

After 15 days, the fish were immobilized, weighed, measured for length and carefully returned to its designated individual aquarium. For every individual fish, the mean of the initial and final weight (g), total weight gain (%), initial and final length (cm), total length gain (%), and specific growth rate (SGR) were recorded and calculated following the given formulae:

I.   Total length gain (TLG)
     Percentage of TLG (%) = $[(L_1 - L_0) \div L_0] \times 100$
     Where, L_0 = Initial mean of total length (cm); L_1 = Final mean of total length (cm)
II. Total weight gain (TWG)
     Percentage of TWG (%) = = $[(W_1 - W_0) \div W_0] \times 100$
     Where; W_0 = Initial mean of body weight (g); W_1 = Final mean of body weight (g)
III. Specific growth rate (SGR)
     Specific weight gain (SGR) (%) = $[(ln$ final body weight $- ln$ initial body weight$) \div$ day$] \times 100$

Statistical analysis
The data were expressed as mean ± SD and analysed by one-way analysis of variance (ANOVA) and Tukey test of multiple comparisons were used for post-hoc statistical evaluation for the fish growth performance with the significant level was set at $P < 0.05$. Statistical analyses were carried out using SPSS (20.0 for windows). All percentage data of total length gain (TLG), total weight gain (TWG) and specific growth rate (SGR) were transformed using Arcsine prior ANOVA.

RESULTS AND DISCUSSION
Salinity tolerance of juveniles AsianSeabass
The results show that Asian Seabass (Figure 1) juveniles were able to survive in all salinity treatments and can tolerate a wide range of salinity (from 0 to 30 ppt). The survival rate of the fish generally is influenced by the ability of the body fluid to tolerate the osmolality of the external environment (Stickeney, 1979). It is reported that Asian seabass is able to accumulate heavy metals such as mercury (Currey et al., 1992), while surviving under various physiological and environmental conditions including varying salinities, high turbidity and temperatures (Job, 2011; Rajaguru, 2002; Yue et al., 2009). This is due to the higher exchange rate, especially on the gill, skin and intestine which are responsible for the water intake (Sarwono, 2004).

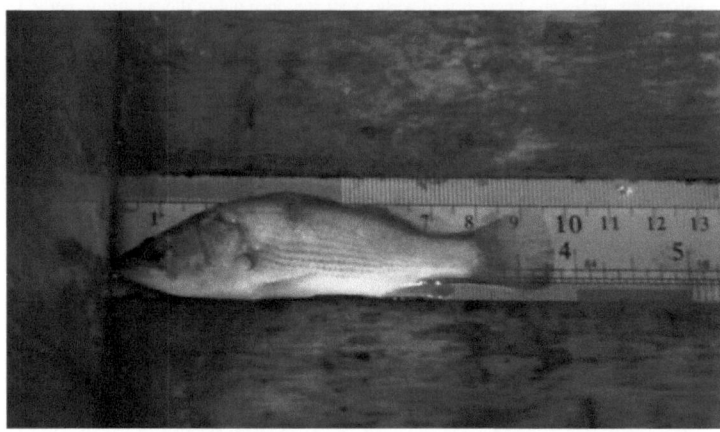

Fig. 1: Asian Seabass (*Lates calcarifer*) of juvenile stage

The observation of fish behavior under different water salinity conditions was also performed within the salinity tolerance trials. The number of fish swimming with abnormal position i.e. nearly 180° body slanting with head pointed downward, (Figure 2) increased gradually from 0 to 10 ppt; with a significantly highest (p<0.05) percentage was observed in 10 ppt with 30%, while there none of the fish swimming with abnormal position in 15 and 20 ppt (Figure 3). However, the percentage of fish swimming with abnormal positions was recorded in 25 and 30 ppt with 10%. This abnormal position suggests that the fish possibly is having problems related to buoyancy. It is possible that the swim bladder may not function properly due to drastic changing water of water quality such as salinity.

Fig. 2: Abnormal swimming position of juveniles Asian Seabass.

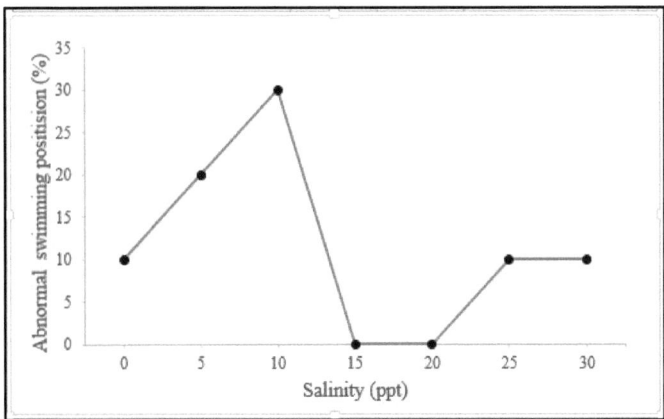

Fig. 3: Percentage of Asian Seabass juvenile with abnormal swimming position in various salinities for 48 hours (n=5).

Effect of different water salinities on growth performance

The mean body length and weight of Asian seabass in all salinities increased within the duration of 15 days (Table 1). The highest mean body length was found in 25ppt salinity; with 10.88±0.12cm and total length gain (TLG) of 8.08 ± 1.81 %. Whereas, the lowest mean of body length was found in 10ppt, with 10.01 ± 0.2 cm of 1.94 ± 0.04 %. TLG was significantly higher (P<0.05) in 0 and 25 ppt.

For total weight gain (TWG), the highest mean body weight was found in 0ppt and with 29.94 ± 14.33 %; while the lowest mean of weight was observed in 15ppt, recorded at 9.58 ± 2.75 %. TWG were significantly higher (P<0.05) in 0, 10 and 25 ppt.

For specific growth rate (SGR), the highest value was obtained at 0 ppt with 1.71 ± 0.74 %/day, while the lowest value of SGR was obtained at 15 ppt with 0.61 ± 0.17 %/day at 15ppt. A significantly higher SGR (P>0.05) was obtained in 0, 10 and 25 ppt.

Table 1: Growth performance parameters, total length gain (TLG), total weight gain (TWG) and specific growth rate (SGR) of juvenile Asian Seabass reared for 15 days at different water salinities (n=5).

Salinity (ppt)	0	5	10	15	20	25	30
TLG (%)	6.16 ± 3.29[a]	4.27 ± 0.31[ab]	1.95 ± 0.12[c]	1.94 ± 0.04[c]	3.13 ± 0.11[b]	8.08 ± 1.81[a]	3.04 ± 0.93[b]
TWG (%)	29.94 ± 14.33[a]	13.79 ± 8.54[b]	24.04 ± 10.13[a]	9.58 ± 2.75[b]	11.10 ± 3.71[b]	26.92 ± 10.21[a]	14.34 ± 8.40[b]

SGR (%)	1.72 ± 0.74[a]	0.85 ± 0.50[b]	1.42 ± 0.54[ab]	0.61 ± 0.17[b]	0.70 ± 0.22[b]	1.58 ± 0.53[a]	0.88 ± 0.49[b]

* Data presented as mean ± standard deviation (SD).[a,b,c] Different superscript indicate a significance different value within a similar row (P<0.05)

Overall, the length gain (TLG), total weight gain (TWG) and specific growth rate (SGR) for Asian seabass is the best in 0 ppt compared to other salinities. The growth performance of fish are affected by genotype-environment interaction such as salinity, photoperiod and temperature (Kikuchi et al., 2007; Zahari et al., 2018) and may also varies according to species, sex, and age (Hepher, 1993; Dutta, 1994).Other than that, factors such as quality and quantity of food, management, and health status also play significant roles. In most fish species, the growth is indeterminate (van Winkle et al., 1997), therefore these factors must be considered when setting up a fish culture to produce fish of the best quality (Boeuf et al., 1999). Some studies reported a better growth rate in intermediary salinity condition such as brackish water as reported in Atlantic salmon, rainbow trout and gilthead sea bream (Boeuf and Payan, 2001) possibly due to hormonal stimulation, slower metabolism, increased feed intake and increased protein digestibility (Kikuchi et al., 2007). However, according to Altinok and Grizzle (2001), some species of juvenile fishes showed inconsistent growth performance when subjected to low salinity due to genetic differences. Salinity deviates the energy available from osmotic regulation to fish growth (Altinok and Grizzle, 2001). However, the relationship between the salinity and growth performance is complex and cannot readily be predicted (Iwama, 1996). For instance, in freshwater fish, the higher salinity, the higher the developmental rate in freshwater fish; contrary to marine fish, the lower water salinity, the higher growth rate reported (Woo & Kell, 1995; Boeuf and Payan,2001).

CONCLUSION
In conclusion, Asian seabass juveniles can tolerate a wide salinity range. However, the juveniles reared at 0 ppt attained the best growth performance as recorded in TWG and SGR, compared to 25 ppt. The results are useful for hatchery management while able to enhance the yield of Asian Seabass, *Lates calcarifer*. Further study on the effect of salinity on the swimming behavior and physiological performance of Asian Seabass is warranted.

ACKNOWLEDGEMENT
The authors would like to thank the Faculty of Fisheries and Food Science, Universiti Malaysia Terengganu for providing the necessary facilities.

REFERENCES
Altinok, I. and Grizzle, J.M. (2001). Effects of brackish water on growth, feed conversion and energy absorption efficiency by juvenile euryhaline and freshwater stenohaline fishes. *Journal of fish Biology*. **59**: 1142-1152.

Amni, R.O., Kawamura, G., Senoo, S. and Ching, F.F. (2015). Effects of different salinities on growth, feeding performance and plasma cortisol level in Hybrid TGGG (Tiger Grouper, *Epinephelus fuscoguttatusx* and Giant Grouper, *Epinephelus lanceolatus*) juveniles. *International Research Journal of Biological Sciences*. **4**: 15-20.

Amornsakun, T., Vo, V.H., Petchsupa, N., Pau, T.M. and Hassan, A.B. (2017). Effects of water salinity on hatching of egg, growth and survival of larvae and fingerlings of snakehead fish, *Channa striatus*. *Songklanakarin Journal Science and Technology*. **39**:137-142.

Boeuf, G., Boujard, D. and Ruyet, J. P. L. (1999). Control of the somatic growth in turbot. *Journal of Fish Biology*. **55**: 128-147.

Boeuf, G. and Payan, P. (2001). How should salinity influence fish growth? *Comparative Biochemistry and Physiology Part C: Toxicology and Pharmacology*. **130**: 411-423.

Boeuf. G. (2009). Acclimatization of aquatic organisms in culture. *Fisheries and Aquaculture-Volume IV*.

In: Encyclopedia of Life Support Systems, EOLSS UNESCO, in press. Pp: 175.

Currey, N.A., Benko, W.I., Yaru, B.T. and Kabi, R. (1992). Determination of heavy metals, arsenic and selenium in Barramundi (*Lates calcarifer*) from Lake Murray, Papua New Guinea. *The Science of the Total Environment*. **125**: 305-320.

Dhert, P., P. Lavens & P. Sorgeloos. (1992). Stress evaluation: a tool for quality control of hatchery-produced shrimp and fish fry. Aquacult. Europe, **17**: 6-10.

Dutta H. (1994). Growth in fishes. *Gerontology (India).* **40**:97-112

Estudillo, C.B., Duray, M.N., Marasigan, E.T. and Emata, A.C. (2000). Salinity tolerance of larvae of the mangrove red snapper (*Lutjanus argentimaculatus*) during ontogeny. *Aquaculture*. **190**: 155-167.

FAO Fisheries statistics (2018). Malaysia Fishery and Aquaculture. FAO Fisheries and Aquaculture Department [online]. Available from: http://www.fao.org/fishery/facp/MYS/en[Accessed on 28th March2018].

Hepher, B. (1993). Growth. In: Hepher B, editor. Nutrition of Pond Fishes. Cambridge: Cambridge University; pp. 163-191

Iwama, G.K. (1996). Growth of salmonids. In Principle of Salmonid Culture (Pennell, W. and Barton, B.A., eds). Amsterdam: Elsevier. Pp. 467-516

Job, S. (2011). Barramundi Aquaculture. *Recent Advances and New Species in Aquaculture*. Pp. 199-229.

Jobling, M. (1995). Fish bioenergetics. *Oceanographic Literature Review*. **9**: 785.

Kikuchi, K., Furuta, T., Ishizuka, H., and Yanagawa, T. (2007). Growth of tiger puffer, *Takifugu rubripes*, at different salinities. *Journal of the World Aquaculture Society*.**38**:427-434.

Kungvankil, P., Tiro Jr, L.B., Pudadera Jr, B.J. and Potesta, I.O. (1985). Training Manual: Biology and Culture of Sea bass (*Lates calcarifer*). Fisheries and Aquaculture Department (FAO) [online]. Available from: http://www.fao.org/docrep/field/003/ac230e/AC230E02.htm#ch2[Accessed on 10th March2018].

Nammalwar, P. and Marichamy, R. (1998). Seabass hatchery. Central Marine Fisheries Research Institute, Kochi. Pp. 149-153.

Nelson, J. (1994). *Fishes of the World*, 3rd edition. John Wiley and Sons, New York.

Rajaguru, S. (2002). Critical thermal maximum of seven estuarine fishes. *Journal of Thermal Biology*. **27**: 125-128.

Sampaio, L.A. and Bianchini, A. (2002). Salinity effects on osmoregulation and growth of the euryhaline flounder *Paralichthys orbignyanus*. *Journal of Experimental Marine Biology and Ecology*. **269**: 187-196.

Sarwono, H.A. (2004). Effect of salinity on osmoregulatory capacity, feed consumption, feed efficiency and growth of juvenile sea bass (*Lates calcarifer* Bloch). KasetsartUniversity.

Shadrin, A.M. and Pavlov, D.S. (2015). Embryonic and larval development of the Asian Seabass *Lates calcarifer* (Pisces: Perciformes: Latidae) under thermostatically controlled conditions. *Izvestiya Akademii Nauk, Seriya Biologicheskaya*. **4**:401-414.

Sharpe, S. (2018). Swim bladder disorder in aquarium fish. The Spruce [online]. Available from: https://www.thespruce.com/swim-bladder-disorder-in-aquarium-fish-1381230[Accessed on 16th April 2018].

Stickney, R.R. (1979). Principles of warmwater aquaculture. *John Wiley and Sons*. New York. Pp. 262-314.

Varsamos, S., Nebel, C. and Charmantier, G. (2005). Ontogeny of osmoregulation in postembryonic fish: A review. *Comparative Biochemistry and Physiology Part A, CBP*. **141**: 401-429.

Van Winkle W, Shuter BJ, Holcomb BD, Jager HI, Tyler JA & Whitaker S (1997). Regulation of energy acquisition and allocation to respiration, growth, and reproduction: simulation model and example using rainbow trout. In: Early Life History and Recruitment in Fish Populations. Chambers RC & Trippel EA (eds.), pp. 103- 137. London, UK: Chapman & Hall

Woo, N. Y. S., & Kell, S. P. (1995). Effect of salinity and nutritional status on growth and metabolism of *Sparus sarba* in a closed seawater system. *Aquaculture*, **135**, 229–238.

Yue, G.H., Zhu, Z.Y., Lo, L.C., Wang, C.M., Lin, G., Feng, F., Pang, H.Y., Li, J., Gong, P., Liu, H.M.,

Tan, J., Chou, R., Lim, H. and Orban, L. (2009). Genetic variation and population structure of Asian seabass (*Lates calcarifer*) in the Asia- Pacific region. *Aquaculture*. **293**: 22-28.

Zahari, Z., Christianus, A., and Ismail, M.F.S. (2018). Effect of stocking density and salinity on the growth and survival of golden Anabas fry. *Survey in Fisheries Sciences*. **4**: 26-37.

Review: Actinomycetes Diversity and Biosynthetic Capabilities of East Coast of Peninsular Malaysia Coastal Water

Zaima Azira Zainal Abidin[1]*, Nurfathiah Abdul Malek

[1]*Dept. of Biotechnology, Kulliyyah of Science, International Islamic University Malaysia*

Corresponding author: zzaima@iium.edu.my

ABSTRACT

Actinomycetes are renowned as an eminent source for antibiotics and wide range of biological compounds. The discovery of streptomycin from *Streptomyces* paved the way for exploration and exploitation of actinomycetes for the discovery of antibiotics and other important compounds. Recognizing the prospective of actinomycetes in natural product discovery, many researchers in Malaysia also took the initiative to participate in actinomycete exploration from local environments. This review summarizes and highlights the research conducted on actinomycete diversity and their biological potential particularly from East Coast of Peninsular Malaysia coastal water namely, Pahang, Terengganu and Kelantan.

Keywords: actinomycetes, diversity, biological activities, coastal water

INTRODUCTION

Actinomycetes are Gram positive, aerobic and filamentous bacteria commonly found in soil. They are renowned for their superior ability in producing secondary metabolites with wide biological activities. Prolific genus *Streptomyces,* for example, accounts for almost 70% of commercially available antibiotics. However, extensive screening of actinomycetes from terrestrial counterpart has led to exhaustion of actinomycetes cultivar and lessened the likelihood of finding novel bioactive secondary metabolites due to the rediscovery of known compounds from previously isolated producers (Lam, 2006; Naikpatil and Rathod, 2011). Hence, exploration of actinomycetes on unexplored and under-explored locations such as extreme environments and marine environment and focusing on rare actinomycete groups may lead to species novelty and eventually to chemical novelty (Goodfellow and Fiedler, 2010; Subramani and Aalbersberg, 2013). Distribution of Malaysian actinomycetes have been studied in mountain range (Lo et al., 2002), rainforest soil (Numata and Nimura, 2003), medicinal plants (Zin et al., 2007), agricultural soils (Jeffrey, 2008), leaf litters (Muramatsu et al., 2011), peat swamp (Jeffrey, 2011), rhizosphere soils (Ting et al., 2009) and compost (Ting et al., 2014). The studies concluded a high diversity of actinomycetes, but with a dominant *Streptomyces* population. Investigation of potential bioactive isolates for enzymatic (Jeffrey et al., 2007; Ting et al., 2014), antibacterial (Jeffrey and Halizah, 2014; Ting et al., 2014) and antifungal (Jeffrey and Halizah, 2014b) activities were also carried out, with promising results that warrant for further investigation. Study on the distribution and biopotential of actinomycetes from Malaysian coastal water environment is still limited particularly in the East Coast of Malaysia making it a prominent source for the isolation and bioprospecting study for drug discovery program. Coastal waters include shelf areas, semi-enclosed and enclosed seas, embayments, estuaries, and wetland areas, often benefit from flows of nutrients from the land and/or also from ocean upwelling which brings nutrient-rich water to the surface providing unique environments for marine bacteria. Moreover, the coastal water environment also experiences various fluctuations of physical factors such as high salinity, high pressure, acidic pH, extreme temperature creating a distinctive environment for marine bacteria including actinomycetes to produce unique and novel secondary metabolites. The East Coast of Peninsular Malaysia encompasses three states namely, Pahang, Terengganu and Kelantan all of which are bordered by the South China Sea in the east. Perhentian and Redang Islands in Terengganu for example are famous for their pristine islands and beaches which present as tourist attractions. East Coast of Peninsular Malaysia holds a great potential as a new resource of highly diverse actinomycetes that can be exploited for natural product discovery. This review discusses the current status of research conducted on actinomycetes diversity and biosynthetic capabilities from East Coast of Peninsular Malaysia coastal water.

Actinomycetes

The name actinomycete derives from the ancient Greek ἀκτίς *(aktís,* 'ray') and μύκης *(múkēs,* 'mushroom or fungus') after the mycelium formation and hyphal tip extension-driven growth. Actinomycetes comprise a large and diverse group of Gram-positive bacteria with high guanine and cytosine ratio (G+C > 55 % mol) in their genome. They are aerobic, slow growing and non-motile which are generally characterized by the formation of thread-like filaments or hyphae (Chaudhary et al., 2013; Goodfellow and Williams, 1983). Actinomycetes play an essential role in nutrient cycling and mineralization of organic matters and in soil, especially the rhizosphere (Murphy, 2007). Taxonomically, actinomycetes are included under the class of Actinobacteria and order of Actinomycetales (Goodfellow and Fiedler, 2010). Actinomycetes comprise 14 suborders, 44 families and over 200 genera with more than 3000 species of bacteria. Members of the order Actinomycetales have been reported as one of the most distributed taxa groups in the domain Bacteria, based on their branching pattern as inferred in the 16S rRNA gene tree (Ventura et al., 2007; Zhi et al., 2009). It should be noted that the expression actinobacteria refers to members of phylum Actinobacteria while the term actinomycetes specifically refers to strains classified under the order Actinomycetales (Goodfellow and Fiedler, 2010). Actinomycetes can be categorized into two major groups: the dominant group and the rare actinomycetes group (Azman et al., 2015). In natural habitat, *Streptomyces* and *Micromonospora* are among the dominant genus of actinomycetes (Genilloud et al. 2011) with more than 900 and 140 species described respectively (www.bacterio.net). On the other hand, genera including *Actinoplanes, Dactylsporangium, Kineosporia, Microbispora* and *Virgosporangium* that have lower isolation rates and are harder to cultivate due to their extremely slow growth are known as rare actinomycetes (Subramani and Sipkema, 2019; Subramani and Aalbersberg, 2013; Tiwari and Gupta, 2013).

Actinomycetes are also known for their economic importance as a result of their great metabolic diversity. They have been commercially exploited for the production of various industrial enzymes including amylase, cellulose, xylanase, proteases and pectinase (Saini et *al.,* 2015). Enzymes produced by actinomycetes not only hold biotechnological importance but can be cost-effective as their production can be carried out by cheap substrates. Actinomycetes also possess the potential for application in soil bioremediation (Timkova et al. 2018), biotransformation and biodegradation of contaminants such as pesticides (Serrano-Gonzalez et al. 2018). They have been the most important sources of bioactive secondary metabolites, many of which have medical importance as antibiotics, antiviral, antiparasitic, antimalarial, antitumor and immunosuppressive agents (Jose and Jha 2016; Demain and Sanchez, 2009). Genus *Streptomyces* alone serves as the most excellent producer, which accounted for more than 10, 400 characterized antimicrobial secondary metabolites followed by *Micromonospora* strains (Berdy, 2012). The capacity of *Streptomyces* strains to produce bioactive compounds especially antibiotics remains incomparable, possibly because of their extra-large DNA complement (Kurtboke, 2012). Rare actinomycetes represented approximately 26% of the antimicrobial compounds with more than 50 rare actinomycete taxa have been reported as the producers of 2,500 antimicrobial compounds (Azman et al. 2015; Subramani and Aalbersberg, 2013). Members of genus *Actinomadura, Actinoplanes, Saccharopolyspora* and *Streptoverticillium* are the most frequent producers among the rare actinomycete groups, each produces hundreds of antibiotics (Subramani and Aalbersberg, 2013),

Selective Isolation of Actinomycetes

One the factors influencing the success of gaining diverse actinomycetes lies in the selective isolation method applied. It is not possible to develop a single procedure for the isolation of different kinds of actinomycetes inhabiting specific environmental samples due to their diverse incubation and growth requirements (Goodfellow, 2010). Accordingly, numerous approaches which include the use of pretreatments procedures and isolation media have been proposed for the isolation of vast actinomycetes taxa groups (Hames and Uzel, 2012). Various pretreatments can be employed to select for different fractions of the community of actinomycetes present in environmental samples (Zainal Abidin et al. 2015; Naikpatil and Rathod, 2011). In general, pretreatment regimes select for target actinomycetes by eliminating the growth of unwanted microorganisms (Goodfellow and Fiedler, 2010; Goodfellow, 2010). Actinomycetes spores are more resistant to desiccation than other bacteria. Thus, air-drying the sediment samples at room temperature will inhibit the colonization of unwanted bacteria that might overrun the isolation plates (Hong *et al.*, 2009). Resistance of actinomycetes propagules to desiccation is commonly associated with their resistance to heat. The main reason for this heat resistance is not clear, but it is apparent that heating prior to inoculation stimulates the germination of actinomycetes spores (Hames and Uzel, 2012). It was reported that many actinomycetes spores (e.g., *Micromonospora* and *Microbispora*), spore vesicles (e.g., *Streptosporangium* and *Dactylsporangium*) and hyphal fragments (e.g., *Rhodococcus*) are more resistant to heat than the Gram-negative prokaryotes (Hames and Uzel, 2012). Heating pretreatment procedures generally lead to a reduction in the ratio of unwanted bacteria to actinomycetes on isolation plates even though the actinomycetes counts may also decrease (Goodfellow, 2010). The use of chemical pretreatments can further enhance their selectivity, as exemplified by the application of benzethonium chloride for the isolation of rare actinomycetes (Bredholt et al., 2008).

Countless isolation media have been designed and proposed for the isolation of actinomycetes. Most of the isolation media were formulated empirically without reference to the nutritional preferences of the target organisms. Most of them have high carbon-to-nitrogen ratio as they contain complex carbon and nitrogen sources (e.g., starch, malt extract, humic acid, casein and xylan) (Hames and Uzel, 2012). These isolation media favor the growth of actinomycetes over the common bacteria which are unable to metabolize the high molecular weight organic polymers. Antimicrobial agents, remarkably actidione, cycloheximide, nystatin and primaricin provide an effective approach to increase the selectivity of isolation media (Liu et al. 2019; Khanna *et al.*, 2011). The use of these antibiotics can be considered as a standard practice to reduce the growth of fungal contaminants. Mimicking the natural habitat is one of the important criteria for the successful isolation of actinomycetes from the natural environment (Goodfellow and Fiedler, 2010). Preparation of isolation media using natural seawater can be crucial for the selective isolation of marine derived actinomycetes (Mincer et al., 2002; Zainal Abidin et al. 2015).

Biosynthetic Genes

A wide range of biologically active compounds with agricultural, medicinal and biotechnological applications are mainly governed by 2 biosynthetic genes remarkably known as nonribosomal polyketide synthases (NRPS) and type-I polyketide synthases (PKS-I) (Ayuso-Sacido and Genilloud, 2005; Gontang et al., 2010). These structurally diverse bioactive metabolites include antibiotics (e.g., erythromycin, nystatin, penicillin and vancomycin), anti-tumor agents (e.g, ansamitocin and bleomycin) and immunosuppressive agents (e.g, rapamycin). Both NRPS and PKS-I biosynthetic pathways have been extensively reported in not only in actinomycetes, but also in cyanobacteria (Fidor et al., 2019) and in filamentous fungi (Theobald et al. 2019). Structurally, both NRPS and PKS-I are multifunctional polypeptides that are encoded by a variable number of modules with multiple enzymatic activities. Each PKS-I module contains 3 domains corresponding to a ketosynthase, acyltransferase and acyl carrier proteins. These domains play an important role in a programmed synthesis of new polyketide chains. Similarly, NRPS modules encoded the activities correspond to the adenylation, condensation and thiolation steps in the recognition and condensation of the substrate. NRPS gene synthesized metabolites which display a remarkable spectrum of activities that were built from individually selected building blocks (Jimenez et al., 2010). Compounds synthesized by NRPS genes are often cyclic in structure and can be distinguished by the presence of non-proteinogenic branched D-amino acids (Miller *et al.*, 2016).

The annotation of biosynthetic gene clusters would complement the bioassay data, enabling manipulation of culturing conditions to stimulate expression of bioactive metabolite (Jimenez et al., 2010). Prediction of bioactive metabolites through genome mining of *Salinispora tropica* leads to isolation and identification of Salinilactam A (Udwary et al., 2007), and likewise, genome mining of two different *Streptomyces* strains that have similar biosynthetic gene cluster leads to the discovery of 3 new polyketides (Banskota *et al.*,

2006). Genome mining of a rare marine actinomycete *Streptosporangium* strain led to the discovery of pentangular polyphenols hexaricins A-C (Tian et al. 2016). Hence, surveying actinomycetes for NRPS and PKS-I biosynthetic genes can be helpful to determine a possible potential of the biological materials (Liu et al. 2019; Zainal Abidin et al. 2018). Positive results in a PCR-based screening not only provide evidence of the production of corresponding metabolites, but also may indicate the existence of further metabolic pathways of secondary metabolite synthesis (Ayuso-Sacido and Genilloud, 2005; Lee et al. 2014). However, the lack of detectable gene fragments does not definitely prove the absence of the respective biosynthetic gene clusters as there are also other metabolites and other biosynthetic pathways that exist as reflected in the actinomycetes genomes (Kouadri et al. 2014; Zainal Abidin et al. 2018).

Diversity and bioactivity of actinomycetes from Pahang, Terengganu and Kelantan
Among all three states, Pahang was the most prolific in terms of research done pertaining to actinomycetes from coastal water environments. One of the hotspot locations for actinomycete research is mangrove forests of Tanjung Lumpur in Kuantan city. Application of selective pretreatments on mangrove sediment samples using a phenol solution (1.5%, 30 min at 30°C) or wet heat in sterilized water (15 min at 50°C) led to the recovery of *Streptomyces, Mycobacterium, Leifsonia, Microbacterium, Sinomonas, Nocardia, Terrabacter, Streptacidiphilus, Micromonospora, Gordonia,* and *Nocardioides* from this location alongside with several possible novel genera and novel species (Lee et al. 2014a). Additionally, detection of PKS-I, PKS-II and NRPS, and evaluation of antimicrobial activity were also conducted on actinomycetes isolated. A number of actinomycetes showed the presence of at least one biosynthetic gene (PKS-I/PKS-II/NRPS) tested and a *Nocardia* species that is closely related to *Nocardia Africana* was found to contain all of the biosynthetic genes (PKS-I, PKS-II, and NRPS). A few *Streptomyces* isolates exhibited antibacterial activity against Methicillin resistant *S. aureus* (MRSA) and one particular *Streptomyces* isolate displayed a broad spectrum of antimicrobial activity which represented novel species named *Streptomyces pluripotens* sp. nov. (Lee et al. 2014b). Accordingly, two novel genera were described namely *Mumia flava* gen. nov. sp. nov (Lee et al. 2014c), and *Monashia flava* gen. nov., sp. nov. (Azman et al. 2016) followed by the description of several novel species - *Microbacterium mangrovi* sp. nov. (Lee et al. 2014d), *Sinomonas humi* sp. nov (Lee et al. 2015), *Streptomyces gilvigriseus* sp. nov (Ser et al. 2015a), *Streptomyces mangrovisoli* sp. nov. (Ser et al. 2015b), *Streptomyces antioxidans* sp. nov. (Ser et al. 2016a), *Streptomyces malaysiense* sp. nov. (Ser et al. 2016b) and *Streptomyces humi* sp. nov. (Zainal et al. 2016). Following the discovery of novel rare actinomycetes from this location, screening on antibacterial, anticancer and neuroprotective activities was conducted on *Microbacterium mangrove, Sinomonas humi* and *Monashia flava* with remarkable findings. Methanolic extracts of *M. mangrove, S. humi* and *M. flava* exhibited bacteriostatic effects while *M. mangrove* extract demonstrated significant neuroprotective properties in oxidative stress and dementia models. Moreover, *M. flava* extract was able to protect the SHSY5Y neuronal cells in the hypoxia model. Additionally, the extracts of *M. mangrovi* and *M. flava* exhibited anticancer effects against human cervical carcinoma cell lines (Ca Ski) (Azman et al. 2017). Further investigation on *Streptomyces gilvigriseus* extract indicated significant antioxidant activity and cytotoxic effect against colon cancer cell lines and this activity might be attributed to cyclic dipeptides present in the extract (Ser et al. 2018).

Similarly, Mohamad et al. (2015) identified 6 *Streptomyces*, 2 *Micromonospora* and 2 *Rhodococcus* with one *Streptomyces* exhibiting broad antimicrobial activity from Tanjung Lumpur including several pathogenic bacteria – *K. pneumoniae, S. thypimurium* and *S. pyogenes*. Bioprospecting program of actinomycetes on 7 locations of Kuantan mangrove forest revealed a highly diverse actinomycetes with high antimicrobial properties. Although genera *Streptomyces* and *Micromonospora* dominated the actinomycetes population, other groups of actinomycetes which belonged to rare actinomycetes were also attained. Members of the rare genera successfully isolated include *Pseudonocardia* sp., *Verrucosispora* sp., *Nocardiopsis* sp., *Actinophytocola* sp., *Dietzia* sp., *Gordonia* sp., *Micrococcus* sp., *Mycobacterium* sp., *Nocardia* sp., *Saccharopolyspora* sp. and *Rhodococcus* sp. Rare actinomycetes strains - *Pseudonocardia* sp., *Nocardiopsis* sp. and *Actinophytocola sp.* also demonstrated antimicrobial activities alongside

71

Streptomyces strains (Abdul Malek et al. 2015, Zainal Abidin et al. 2018). Besides *Streptomyces* and *Micromonospora* isolates displaying the presence of PKS-I and/or NRPS genes in them, several rare actinomycetes - *Actinophytocola, Gordonia, Pseudonocardia, Rhodococcus* and *Verrucosispora* showed similar observation too.

One isolate of particular interest, *Actinophytocola* sp. K4–08 which was recovered through dry heat pretreatment 120°C, 60 min on ISP4 medium. This actinomycete was closely related to *A. sediminis* (99% similarity) that was previously found in the deep sea sediment of South China Sea. This isolate possessed both NRPS and PKS-I biosynthetic genes and displayed promising antimicrobial activity against the test organisms. Evaluation on the antimicrobial activities and biosynthetic capabilities of genus *Actinophytocola* has never been reported before making this isolate as a promising candidate to be exploited for natural product discovery. Additionally, several actinomycetes were found to produce coloured diffusible pigment (Figure 1). Production of diffusible pigment is usually related to the melanin release into the medium and pigments play significant roles in the survival and growth of actinomycetes (Parungao et al. 2007). Occasionally other pigment colour was also reported such as yellow, green and blue and sometimes these pigments are exhibiting antimicrobial activities. Besides brown and black as the common diffusible pigment obtained from actinomycetes, diffusible pigments of blue, orange, pink, purple and yellow were reported in their study. Moreover, ethyl acetate extract of the purple pigment possessed strong inhibitory activity against *B. subtilis, S. aureus* and *S. marcescens.*

Next location in Pahang is Tioman Island which is surrounded by the South China Sea and considered to be an untapped source of rare marine actinomycetes. Sabaratnam et al. (2008) reported diverse actinomycetes isolated from marine sponges collected in Tioman Island and putatively identified selected isolates as *Actinoplanes* spp., *Micromonospora* spp., *Nocardia* spp., *Polymorphospora* spp., *Pseudonocardia* spp., *Rhodococcus* spp., *Saccharomonospora* spp., *Salinispora* spp., *Sprilliplanes* spp. and *Verrucosispora* spp. In a more recent study by Ng and Tan (2018) on marine sediment collected from the Pirate Reef, Tioman Island, analyses of the 16S rRNA gene sequences indicated close relationships to members of 18 genera: *Actinomadura, Agromyces, Jishengella, Marinactinospora, Micromonospora, Mycobacterium, Nocardia, Nocardiopsis, Nonomuraea, Plantactinospora, Pseudonocardia, Rhodococcus, Saccharomonospora, Saccharopolyspora, Salinispora, Streptomyces,* and *Streptosporangium.* Moreover, almost half of the isolates recovered were *Streptomyces* spp. (47.97%) and *Salinispora* spp. (23.58%), respectively. This was followed by description of the novel genus *Marinitenerispora sediminis* gen. nov., sp. nov and this bacterium also possessed inhibitory activity against B. *subtilis, S. aureus* and *E. coli* (Ng et al. 2019). Another actinomycetes research by Zainal Abidin (2013) reported the occurrence of *Streptomyces* and *Salinispora* isolates from marine sediment of Tioman Island (Figure 2). *Streptomyces* isolates displaying strong antimicrobial activity and *Salinispora* isolate exhibited strong antibacterial activity against pathogenic MRSA. One particular *Streptomyces* isolate was able to tolerate up to 12% NaCl indicating its adaptation to marine environment. Tioman Island seems to be hotspot for *Salinispora* strains as demonstrated by several studies indicating the presence of this obligate marine actinomycetes as indigenous actinomycete in the marine sediment of Tioman Island. Another location in Pahang is Cherating in which Ariffin et al. (2017) successfully isolated *Streptomyces* from the mangrove area located here. Extensive studies on actinomycetes in Pahang localities coupled with the recovery of rare actinomycetes and the description of novel genus and species further exemplify the true potential of Pahang coastal water as new resources of actinomycetes with biosynthetic abilities.

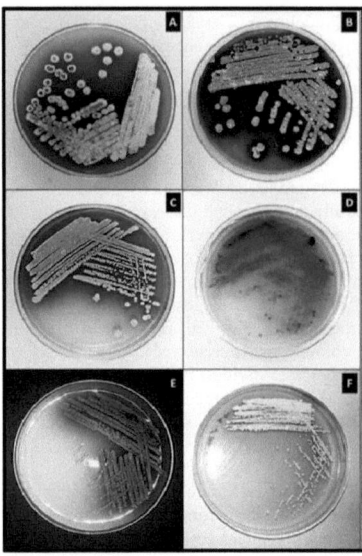

Fig. 1: Coloured diffusible pigment from actinomycetes of Kuantan Mangrove Forest

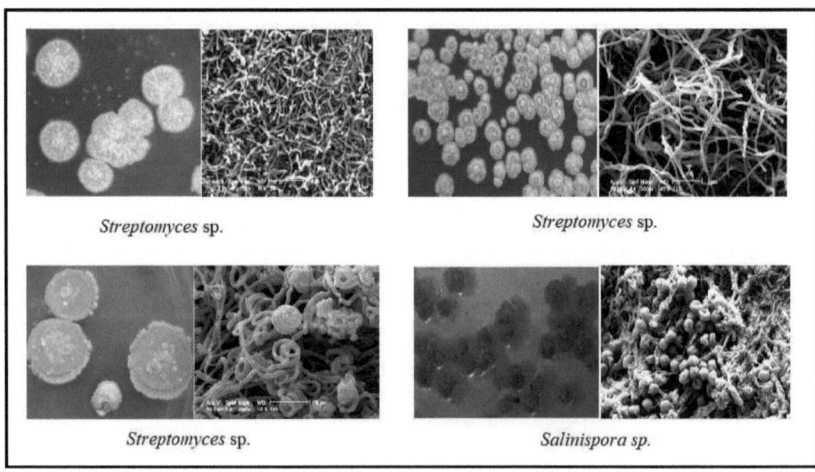

Fig. 2: Colonies morphologies and SEM micrographs of actinomycetes from Tioman Island

However, few studies were conducted on actinomycetes from coastal water Terengganu and Kelantan. Ariffin et al. (2017) isolated a total of 11 actinomycetes isolates from Chendering beach in Terengganu and

7 actinomycetes from mangrove sediment in Tok Bali beach, Kelantan though their identities were not determined. Another location in Terengganu is Bidong Island. This island previously was a refugee camp for Vietnamese and it opened to tourists after all refugees were repatriated to Vietnam. Recently, cultivable bacteria associated with different species of sea sponges collected adjacent to Bidong Island recovered *Brevibacterium* and *Kytococcus* among the bacteria population identified (Tan et al. 2018). Next, a study focusing on bacteria associated with mucus of *Acropora cervicornis* coral also in Bidong Island recovered *Actinomyces, Micrococcus varians, Micrococcus roseus* and *Micrococcus* sp. alongside with other groups of bacteria (Kalimuthu et al. 2007). Certainly, there are other research directed in actinomycetes isolation and diversity in the states of Kelantan and Terengganu but yet to be reported. Unquestionably, coastal waters located in Kelantan and Terengganu hold the prospective to be new resources of actinomycetes with potentially novel compounds just waiting to be explored and discovered. Table 1 summarized the diversity of actinomycetes and their bioactivities according to each state – Pahang, Terengganu and Kelantan. Indeed, coastal waters of the East Coast of Peninsular Malaysia do hold the potential to be explored as a new resource of actinomycetes. Perhaps, concerted and strategic effort by various research groups especially on bioprospecting of actinomycetes on these locations may yield novel strains and lead to the discovery of unique bioactive compounds.

Table 1: Summary of actinomycetes from coastal water of Pahang, Terengganu and Kelantan

State	Genus	Bioactivity	Reference
	Tanjung Lumpur *Streptomyces, Mycobacterium, Leifsonia, Microbacterium, Sinomonas, Nocardia, Terrabacter, Streptacidiphilus, Micromonospora, Rhodococcus, Gordonia, Nocardioides, Mumia flava, Monashia flava*	Antibacterial, anticancer, antioxidant, neuroprotective activities	Lee et al. (2014a); Lee et al. (2014b); Lee et al. (2014c); Lee et al. (2014d); Azman et al. (2016); Mohamad et al. (2015); Ser et al. (2015a); Ser et al. (2015b); Ser et al. (2016a); Ser et al. (2016b); Zainal Abidin et al. (2016); Azman et al. (2017); Ser et al. (2018)
Pahang	**Kuantan Mangrove Forest** *Pseudonocardia, Verrucosispora, Nocardiopsis, Actinophytocola, Dietzia, Gordonia, Micrococcus, Mycobacterium, Nocardia, Saccharopolyspora, Rhodococcus, Pseudonocardia, Nocardiopsis, Actinophytocola*	Antimicrobial	Abdul Malek et al. (2015); Zainal Abidin et al. (2018)
	Tioman Island *Actinoplanes, Micromonospora, Nocardia, Polymorphospora, Pseudonocardia,*		

	Rhodococcus, Saccharomonospora, Salinispora, Sprilliplanes, Verrucosispora, Actinomadura, Agromyces, Jishengella, Marinactinospora, Mycobacterium, Nocardiopsis, Nonomuraea, Plantactinospora, Saccharopolyspora, Streptosporangium, Streptomyces, Marinitenerispora sediminis	Antimicrobial	Sabaratnam et al. (2008); Zainal Abidin (2013); Ng & Tan (2018); Ng et al. (2019)
	Cherating		
	Streptomyces	Antibacterial	Ariffin et al. (2017)
	Bidong Island		
	Brevibacterium, Kytococcus, Actinomyces, Micrococcus	Not determined	Kalimuthu et al. (2007); Tan et al. (2018)
Terengganu			
	Chendering	Antibacterial	Ariffin et al. (2017)
	Unknown		
Kelantan	**Tok Bali Beach**	Not determined	Ariffin et al. (2017)
	Unknown		

CONCLUSION

Description of novel genus and novel species from coastal water from East Coast of Peninsular Malaysia demonstrated the perspective of actinomycetes from coastal waters of Pahang, Terengganu and Kelantan and the possible potential application in natural product discovery. Though studies on actinomycetes from these states are still lacking, nevertheless, these locations hold the prospective to be hotspots for novel actinomycetes and novel compounds. Research on actinomycetes should go beyond than diversity and biological screening activities but to attempt in purification and structure elucidation of bioactive compounds as well to embark new avenues such as genome mining, next generation sequencing (NGS), metabolomics and proteomics to disclose cryptic biosynthetic pathways in secondary metabolites production.

REFERENCES

Abdul Malek, N., Zainuddin, Z., Chowdhury, A.J.K., Zainal Abidin, Z.A. (2015). Diversity and antimicrobial activity of mangrove soil actinomycetes isolated from Tanjung Lumpur, Kuantan, *Jurnal Teknologi, 77*(25), 37–43.

Ariffin, S., Abdullah, M.F.F., Mohamad, S.A.S. (2017). Identification and Antimicrobial Properties of Malaysian Mangrove Actinomycetes, *Int. J. on Advanced Science Engineering Information Technology, 7*(1), 71-77.

Ayuso-Sacido, A. and Genilloud, O. (2005). New PCR Primers for the screening of NRPS and PKS-I systems in actinomycetes: detection and distribution of these biosynthetic gene sequences in major taxonomic groups, *Microbial Ecology*, 49, 10–24.

Azman, A. S., Iekhsan, O., Velu, S. S., Chan, K. G. and Lee, L. H. (2015). Mangrove rare actinobacteria: taxonomy, natural compound, and discovery of bioactivity, *Frontiers in Microbiology*, 6, 85601–85615.

Azman, A. S., Zainal, N., Ab Mutalib, N.S., W.F. Chan, K. G. and Lee, L.H. (2016). *Monashia flava* gen. nov., sp. nov., an actinobacterium of the family Intrasporangiaceae, *Int J Syst Evol Microbiol*, 66, 554–561.

Azman, A. S., Othman, I., Fang, C.M., Chan, K. G., Goh, B.H., Lee, L.H. 2017. Antibacterial, Anticancer and Neuroprotective Activities of Rare Actinobacteria from Mangrove Forest Soils, *Indian J Microbiol*, 57(2),177–187.

Berdy, J. (2005). Bioactive microbial metabolites, *The Journal of Antibiotics*, 58,1–26.

Bredholt, H., Fjaervik, E., Johnsen, G. and Zotchev, S. B. (2008). Actinomycetes from sediments in Trondheim Fjord, Norway: diversity and biological activity, *Marine Drugs*, 6, 12–24.

Chaudhary, H. S., Soni, B., Shrivastava, A. R. and Shrivastava, S. (2013). Diversity and versatility of actinomycetes and its role in antibiotic production, *Journal of Applied Pharmaceutical Science*, 3: S83–S94.

Demain, A. L. and Sanchez, S. (2009). Microbial drug discovery: 80 years of progress. *The Journal of Antibiotics*, 62: 5–16.

Fidor, A., Konkel, R. and Mazur-Marzec, H. (2019). Bioactive Peptides Produced by Cyanobacteria of the Genus Nostoc: A Review, *Mar. Drugs*, 17, 561 doi:10.3390/md17100561

Genilloud, O., Gonzalez, I., Salazar, O., Martin, J., Tormo, J. R. and Vicente, F. (2011). Current approaches to exploit actinomycetes as a source of natural products, *Journal of Industrial Microbiology and Biotechnology*, 38, 375–389.

Gontang, A. E., Gaudencio, S. P., Fenical, W. and Jensen, P. R. (2010). Sequence-based analysis of secondary-metabolite biosynthesis in marine actinobacteria, *Applied and Environmental Microbiology*, 76, 2487–2499.

Goodfellow, M. (2010). Selective Isolation of Actinobacteria. *In* Baltz, D. R. H. and Davies, J. (Eds.), *Manual of Industrial Microbiology and Biotechnology*. (3rd ed., pp. 13–27). Washington DC: ASM Press.

Goodfellow, M. and Fiedler, H. P. (2010). A guide to successful bioprospecting: informed by actinobacterial systematics, *Antonie Van Leeuwenhoek*, 98, 119–142.

Goodfellow, M. and Williams, S. T. (1983). Ecology of actinomycetes, *Annual Review of Microbiology*, 37, 189–216.

Hames, E. E. and Uzel, A. (2012). Isolation strategies of marine-derived actinomycetes from sponge and sediment samples, *Journal of Microbiological Methods*, 88, 342–347.

Hong, K., Gao, A. H., Xie, Q. Y., Gao, H., Zhuang, L., Lin, H. P., Yu, H. P., Li, J.., Yao, X. S., Goodfellow, M. and Ruan, J. S. (2009). Actinomycetes for marine drug discovery isolated from mangrove soils and plants in China, *Marine Drugs*, 7, 24–44.

Jeffrey, L. S. H., Sahilah, A. M., Son, R. and Tosiah, S. (2007). Isolation and screening of actinomycetes from Malaysian soil for their enzymatic and antimicrobial activities, *Journal of Tropical Agriculture and Food Science*, 1, 159–164.

Jeffrey, L. S. H. (2008). Isolation, characterization and identification of actinomycetes from agriculture soils at Semongok, Sarawak. *African Journal of Biotechnology*, 7, 3697–3702.

Jeffrey, L. S. H. (2011). Presecreening of bioactivities from actinomycetes isolated from forest peat soil of Sarawak, *Journal of Tropical Agriculture and Food Science*, 39, 245–253.

Jeffrey, L. S. H. and Halizah, H. (2014). Biological active compounds from actinomycetes isolated from soil of Langkawi Island, Malaysia, *African Journal of Biotechnology*, 13, 4523–4528.

Jimenez, J. T., Sturdikova, M. and Sturdik, E. (2010). Bioactive marine and terrestrial polyketide and peptide secondary metabolites and perspectives of their biotechnological production, *Acta Chimica Slovaca*, 3, 103–119.

Jose, P.A. and Jha, B. (2016). New Dimensions of Research on Actinomycetes: Quest for Next Generation Antibiotics, *Front. Microbiol.* 7:1295. doi: 10.3389/fmicb.2016.01295.

Kalimutho, M., Ahmad, A. and Kassim, Z. (2007). Isolation, Characterization and Identification of Bacteria associated with Mucus of *Acropora cervicornis* Coral from Bidong Island, Terengganu, Malaysia, *Malaysian Journal of Science* 26 (2), 27 – 39.

Khanna, M., Solanki, R. and Lal, R. (2011). Selective isolation of rare actinomycetes producing novel antimicrobial compounds, *International Journal of Advanced Biotechnology and Research*, 2, 357–375.

Kouadri, F.; Al-Aboudi, A., and Jorani, H.K., (2014). Antimicrobial activity of Streptomyces sp. isolated from the Gulf of Aqaba-Jordan and screening for NRPS, PKS-I and PKS-II genes, *African Journal of Biotechnology,* 13(34), 3505–3515

Kurtboke, D. I. (2012). Biodiscovery from rare actinomycetes: an eco-taxonomical perspective, *Applied Microbiology and Biotechnology*, 93, 1843–1852.

Lam, K. S. (2006). Discovery of novel metabolites from marine actinomycetes, *Current Opinion in Microbiology*, 9, 245–251.

Lee, L. H., Nurullhudda, Z. Adzzie-Shazleen, A., Eng, S. K., Goh, B. H., Yin, W. F., Nurul-Syakima, A. M. and Chan, K. G. (2014a). Diversity and antimicrobial activities of actinobacteria isolated from tropical mangrove sediments in Malaysia, *The Scientific World Journal*, 10, 1–14.

Lee, L. H., Nurullhudda, Z. Adzzie-Shazleen, A., Eng, S. K., Nurul-Syakima, A. M., Yin, W.F. and Chan, K. G. (2014b). *Streptomyces pluripotens* sp. nov., a bacteriocin-producing streptomycete that inhibits meticillin-resistant *Staphylococcus aureus*, *Int J Syst Evol Microbiol*, 64, 3297–3306.

Lee, L. H., Nurullhudda, Z. Adzzie-Shazleen, A., Nurul-Syakima, A. M., Hong, K. and Chan, K. G. (2014c). *Mumia flava* gen. nov., sp. nov., an actinobacterium of the family Nocardioidaceae, *Int J Syst Evol Microbiol* 64: 1461–1467.

Lee, L. H., Adzzie-Shazleen, A., Nurullhudda, Z. Eng, S.K., Nurul-Syakima, A. M., Yin, W.F. and Chan, K. G. (2014). *Microbacterium mangrovi* sp. nov., an amylolytic actinobacterium isolated from mangrove forest soil, *Int J Syst Evol Microbiol* 64, 3513–3519.

Lee, L. H., Adzzie-Shazleen, A., Nurullhudda, Z., Yin, W.F., Nurul-Syakima, A. M., and Chan, K. G. (2015). *Sinomonas humi* sp. nov., an amylolytic actinobacterium isolated from mangrove forest soil, *Int J Syst Evol Microbiol*, 65, 996–1002.

Liu, T., Wu, S., Zhang, R., Wang, D., Chen, J. and Zhao, J. (2019). Diversity and antimicrobial potential of Actinobacteria isolated from diverse marine sponges along the Beibu Gulf of the South China Sea, *FEMS Microbiology Ecology*, 95(7) doi: 10.1093/femsec/fiz089

Lo, C. W., Lai, N. S., Cheah, H. Y., Wong, N. K. I. and Ho, C. C. (2002). Actinomycetes isolated from soil samples from the Crocker range Sabah, *ASEAN Review of Biodiversity and Environmental Conversation*, 9, 1–7.

Miller, B.R., Drake, E.J, Shi, C., Aldrich, C.C. and Gulick, A.M. (2016). Structures of a Nonribosomal Peptide Synthetase Module Bound to MbtH-like Proteins Support a Highly Dynamic Domain Architecture, *The Journal of Biological Chemistry* 291(43), 22559 –22571.

Mohamad, N.H., Chowdhury, A.J.K. and Zainal Abidin, Z.A. (2015). Selective isolation of Actinomycetes from mangrove sediment of Tanjung Lumpur, Kuantan, Malaysia, *Malaysian Journal of Microbiology*, 11(2), 144-155.

Muramatsu, H., Murakami, R., Ibrahim, Z. H., Murakami, K., Shahab, N. and Nagai, K. (2011). Phylogenetic diversity of acidophilic actinomycetes from Malaysia, *The Journal of Antibiotics*, 64, 621–624.

Murphy, D. V., Stockdale, E. A., Brookes, P. C. and Goulding, K. W. T. (2007). Impact of microorganisms on chemical transformations in soil. *In* Abbot, L. K. and Murphy, D. V. (Eds.). *A Key to Sustainable Land Use in Agriculture*. (1st ed., pp. 37–59). New York: Springer.

Naikpatil, S. V. and Rathod, J. L. (2011). Selective isolation and antimicrobial activity of rare actinomycetes from mangrove sediment of Karwar, *Journal of Ecobiotechnology*, 3, 48–53.

Ng, Z.Y. and Tan, G.Y.A. 2018. Selective isolation and characterisation of novel members of the family Nocardiopsaceae and other actinobacteria from a marine sediment of Tioman Island, *Antonie van Leeuwenhoek* 111, 727–742.

Ng, Z.Y., Fang, B.Z., Li, W.J.and Tan, G.Y.A. (2019). *Marinitenerispora sediminis* gen. nov., sp. nov., a member of the family Nocardiopsaceae isolated from marine sediment *Int J Syst Evol Microbiol, 69,* 3031–3040.

Numata, K. and Nimura, S. (2003). Access to soil actinomycetes in Malaysian tropical rain forests, *Actinomycetologica*, 17, 54–56.

Parungao, M. M., Maceda, E. B. G. and Vilano, M. A. F. (2007). Screening of antibiotic-producing actinomycetes from marine, brackish and terrestrial sediments of Samal Island, Philippines, *Journal of Research in Science, Computing and Engineering*, 4, 29–38.

Sabaratnam, V., Christabel, L.J., Thong, K.L., Tan, G.Y.A., Affendi, Y.A. (2008). *Sponges of Tioman and their actinomycetes inhabitants.* In: Natural history of the Pulau Tioman Group of Islands. IOES monograph series. University of Malaya, Kuala Lumpur, pp. 35-41. ISBN 9789839576351

Saini, A., Aggarwal, N.K., Sharma, A. and Yadav, A. (2015). Actinomycetes: A Source of Lignocellulolytic Enzymes, *Enzyme Research*, 20, 1-15.

Ser, H.L., Zainal, N. Palanisamy, U.D., Goh, B.H., Yin, W.F., Chan, K.G. Lee, L.H. (2015a). *Streptomyces gilvigriseus* sp. nov., a novel actinobacterium isolated from mangrove forest soil, *Antonie van Leeuwenhoek,* 107,1369–1378.

Ser, H.L., Palanisamy U.D., Yin W.F., Abd Malek S.N., Chan K.G., Goh B.H. and Lee L.H. (2015b). Presence of antioxidative agent, Pyrrolo[1,2-a] pyrazine-1,4-dione, hexahydro- in newly isolated *Streptomyces mangrovisoli* sp. nov., *Front. Microbiol.* 6, 854. doi: 10.3389/fmicb.2015.00854

Ser, H.L., Tan, L.T.H., Palanisamy, U.D., Abd Malek, S.N., Yin, W.F., Chan, K.G., Goh, B.H. and Lee, L.H. (2016a) *Streptomyces antioxidans* sp. nov., a Novel Mangrove Soil Actinobacterium with Antioxidative and Neuroprotective Potentials, *Front. Microbiol.* 7:899. doi: 10.3389/fmicb.2016.00899

Ser, H.L., Palanisamy, U.D., Yin, W.F., Chan, K.G., Goh, B.H. and Lee, L.H. (2016b). *Streptomyces malaysiense* sp. nov.: A novel Malaysian mangrove soil actinobacterium with antioxidative activity and cytotoxic potential against human cancer cell lines, *Scientific Reports* 6, 24247 doi: 10.1038/srep24247

Ser, H.L., Yin, W.F., Chan, K.G, Goh, B.H., Lee, L.H. 2018. Antioxidant and cytotoxic potentials of *Streptomyces gilvigriseus* MUSC 26T isolated from mangrove soil in Malaysia, *Prog Microbes Mol Biol* 1(1), a0000002.

Serrano-Gonzalez, M.Y., Chandra, R., Castillo-Zacarias, C., Robledo-Padilla, F., Rostro-Alanis, M.J., Parra-Saldivar, R. (2018). Biotransformation and degradation of 2,4,6-trinitrotoluene by microbial metabolism and their interaction, *Defence Technology,* 14, 151-164.

Subramani, R. and Sipkema, D. (2019). Marine Rare Actinomycetes: A Promising Source of Structurally Diverse and Unique Novel Natural Products, *Marine Drugs*, 17, 249; doi:10.3390/md17050249

Subramani, R. and Aalsberg, W. (2013). Culturable rare actinomycetes: diversity, isolation and marine natural product discovery, *Applied Microbiology and Biotechnology*, 97, 9291–9321.

Tan, S.M.A., Amirul, A.A., Saidin, J. and Bhubalan, K. (2018). Identification of Cultivable Bacteria from Tropical Marine Sponges and Their Biotechnological Potentials, *Tropical Life Sciences Research*, 29(2), 187–199.

Theobald, S., Vesth, T.C. and Andersen, M.R. (2019). Genus level analysis of PKS-NRPS and NRPS-PKS hybrids reveals their origin in Aspergilli, *BMC Genomics,* 20,847.

Tian, J., Chen, H., Guo, Z., Liu, N., Li, J., Huang, Y., Xiang, W. and Chen, Y. (2016). Discovery of pentangular polyphenols hexaricins A–C from marine *Streptosporangium* sp. CGMCC 4.7309 by genome mining, *Appl Microbiol Biotechnol,* 100, 4189–4199.

Timková, I., Jana Sedláková-Kaduková, J. and Prista, P (2018). Biosorption and Bioaccumulation Abilities of Actinomycetes/Streptomycetes Isolated from Metal Contaminated Sites, *Separations*, 5(54); doi:10.3390/separations5040054

Ting, A. S. Y., Tan, S. H. and Wai, M. K. (2009). Isolation and characterization of actinobacteria with antibacterial activity from soil and rhizosphere soil. *Australian Journal of Basic and Applied Sciences*, 3, 4053–4059.

Ting, A. S. Y., Hermanto, A. and Peh, K. L. (2014). Indigenous actinomycetes from empty fruit bunch compost of oil palm: evaluation on enzymatic and antagonistic properties, *Biocatalysis and Agricultural Biotechnology*, 3, 310–315.

Tiwari, K. and Gupta, R. K. (2013). Diversity and isolation of rare actinomycetes: an overview, *Clinical Reviews in Microbiology*, 39, 256–294.

Ventura, M., Chancaya, C., Tauch, A., Chandra, G., Fitzgerald, G. F., Chater, K. F. and Sinderen, D. V. (2007). Genomics of actinobacteria: tracing the evolutionary history of an ancient phylum, *Microbiology and Molecular Biology Reviews*, 71, 495–548.

Zainal, N., Ser, H.L., Yin, W.F., Tee, K.K., Lee, L.H., Chan, K.G. 2016. *Streptomyces humi* sp. nov., an actinobacterium isolated from soil of a mangrove forest, *Antonie van Leeuwenhoek,* 109, 467–474.

Zainal Abidin, Z.A. Actinomycetes Diversity and Characterisation of Bioactive Compounds of *Streptomyces* from Malaysian Marine Environment. PhD Thesis. Universiti Kebangsaan Malaysia. 2013. 247p.

Zainal Abidin, Z.A., Abdul Malek, N., Zainuddin, Z., Chowdhury, A.J.K. (2015). Selective isolation and antagonistic activity of actinomycetes from mangrove forest of Pahang, Malaysia, *Frontiers in Life Science*, 9(1), 24-31

Zainal Abidin, Z.A., Chowdhury, A.J.K., Abdul Malek, N., Zainuddin, Z. (2018). Diversity, Antimicrobial Capabilities, and Biosynthetic Potential of Mangrove Actinomycetes from Coastal Waters in Pahang, Malaysia, *Journal of Coastal Research* 82, 174-179.

Zhi, X. Y., Li, W. J. and Stackebrandt, E. (2009). An update of the structure and 16S rRNA gene sequence-based definition of higher ranks of the class *Actinobacteria*, with the proposal of two new suborders and four new families and emended description of the existing higher taxa, *International Journal of Systematic and Evolutionary Microbiology*, 59, 589–608.

Zin, N. M., Sarmin, N. I. M., Ghadin, N., Basri, D. F., Sidik, N. M., Hess, W. M. and Strobel, G. A. (2007). Bioactive endophytic streptomycetes from the Malay Peninsula, *FEMS Microbiology Letters*, 274, 83–88.

Climate Change and Coastal Defense in Malaysia: A Review

Muhammad Zahir Ramli[1]*, Muhammad Adil Ramzi[2], Muhamad Syafiq Safwan[2], Nur Adawiyah Isa[2], Minhalina Ahmad[2], Nur Azierah Samsu Bahari[2], Kamaruzzaman, B.Y[1]

[1]*Department of Marine Science, Kulliyyah of Science, International Islamic University Malaysia, 25200 Kuantan, Pahang, Malaysia*
[2]*Institute of Oceanography & Maritime Studies, Kulliyyah of Science, International Islamic University Malaysia, 25200 Kuantan, Malaysia*
Corresponding author: mzbr@iium.edu.my

ABSTRACT

Coastal zones around the world are facing increased numbers of populations through rapid development and expansion for residential, industry and tourism areas. There are approximately 50% of the global population that inhabit coastal areas. With the current climate change, coastal zones are exposed to sea level rise and flooding which could bring catastrophe to low-lying regions. Many countries have developed mitigation and adaptation plans where most of the approaches involve in the alteration of natural coastline through the construction of coastal defenses. There are many key strategies in the implementation of coastal defense with the aim to reduce or minimize the impact towards the shoreline. This review provides insight on different approaches of coastal defenses in Malaysia, specifically focusing on erosion or flooding, morphological conditions, and land use. This article also highlights the improvement needed to withstand the impact of sea level rise. This review will benefit researchers who would like to explore the key parameter in structure design of coastal defense.

Keywords: Climate Change, Coastal Defense, Erosion, Overtopping, Coastal Management.

INTRODUCTION

Coastal zones are vulnerable environments that continuously receive harmful threats. Those threats are typically resulted from the massive development and rapid urbanization of the coastal areas as well as natural-based phenomena such as climate change and sea level rise. With that, multiple initiatives have been taken in order to overcome the coastal zone related issues particularly involving the shoreline erosion problems. Numerous coastal protection structures have been developed along the affected coastlines in Malaysia. Such structures involve both soft and hard engineering structures. Primarily, by constructing the coastal protection structures, erosion and flooding of high value coastlines can be prevented and reduced, beaches and reclaimed land can be stabilized as well as the amenity value of the coast can be enhanced. At the global scale, the proliferation of artificial coastal protection structures in the marine environment is mainly related to climate change adaptations which simultaneously aims to keep up with the increasing commercial and recreational uses of the coastal zones.

However, without appropriate plan and design prior to the construction of the coastal protection structures as well as lack of maintenance, numerous problems might potentially arise at certain periods of time after the construction. One of the major problems include the interruption of the littoral sediment transports which eventually might lead to the sediment deposition process. Besides that, improper design might contribute to the collapse of the coastal protection structures. Above all, these problems indicate the failure of the structures hence exert greater challenges to the coastal management. Therefore, this review intends to discuss several components which include the major threats to the coastal zones, the coastal protection structures that have been constructed in Malaysia, the challenges to the coastal protection structures as well as certain suggestions to be applied in order to overcome the existing challenges.

Major threats to Coastal Zones

Coastal zones experience tremendous changes due to the introduction of both natural and anthropogenic pressures. These pressures have directly and indirectly disrupted the stability of shorelines. Shoreline erosion is one the major threats. The imbalance between the supply and export of materials which are mainly dominated by sediments to and from a coastal area can be recognized as the shoreline erosion (Najib, Ab Ghani, Abdullah & Ahmad, 2017). An eroded coastline can be commonly detected through the landward displacement of the shoreline. Based on the National Coastal Erosion Study 1984, around 29% or 1,380 km of Malaysian coastlines experienced erosional problems whereby 52% of them occurred in Peninsular Malaysia (Ministry of Natural Resources and Environment, 2009). Urbanization along the coastal zones is one of the major contributors. The coastal zones in Malaysia have become the center of urban and rural economic activities whereby up to 70% of the Malaysian population live within the coastal areas (Najib et. al., 2017).

Apart from that, natural components such as wind, waves, tides as well as currents are also included among the contributors to the coastal erosion. In certain months throughout a year, Peninsular Malaysia particularly are prone to wind related phenomena which are known as monsoon seasons. These phenomena subsequently worsen the coastal erosion problems. Study shows that there is an increment in the coastal erosion cases in Peninsular Malaysia from 2013 to 2017 (Yanalagaran, et al. 2019). Generally, a significant correlation can be observed between the average wind speeds and the number of erosion cases (Figure 1). It is found that in the month of February and December, the highest cases of coastal erosion are aligned with the fastest average wind speed. These two months fall under the duration of Northeast monsoon season which is between November to March. On the other hand, during the Southwest monsoon which is between May to September, the least number of erosion cases with some fluctuations are observed. In other words, the occurrence of the Northeast monsoon exerts greater impacts to the coastal erosion in Peninsular Malaysia than the Southwest monsoon.

Moreover, out of 14 states in the Peninsular Malaysia, nine of them suffer coastal erosion problems. Such states include Johor, Melaka, Negeri Sembilan, Kelantan, Pahang, Pulau Pinang, Perak, Selangor and Terengganu (Table 1). In Malaysia, based on the National Coastal Erosion Study 2015, up to 44 beaches experienced erosion as a whole and have been classified under the Category 1 which is referred to as critical cases (Department of Irrigation and Drainage Malaysia, 2015).

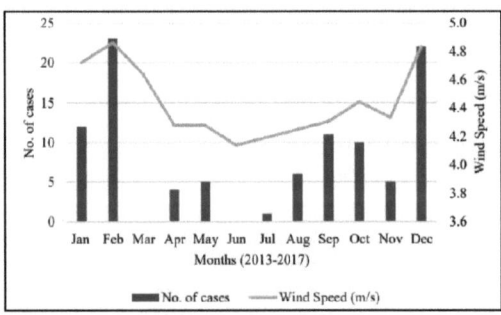

Fig. 1: Graphs of wind speed and number of coastal erosion cases in Peninsular Malaysia (Yanalagran et al., 2019)

Table 1: Length of Eroded Coastline in Different Beaches in Malaysia

State	Beach	Length of Eroded Coastline (m)
Kedah	Pantai Pasir Hitam	345.5
	Kampung Penarek	134.1
	Kampung Padang Salin	649.5
Pulau Pinang	Persiaran Bayan Indah	1138.4
	Taman Molek	438.7
	Persiaran Bayan Mutiara	610
	Kampung Benggali	263
	Kampung Kuala Muda	598.1
	West of Kampung Benggali	828.1
	Kampung Permatang Rawa	1678.1
Perak	Kuala Kurau	1861
Selangor	Kampung Batu Laut	1384.9
	Pantai Jeram – Pantai Remis	3438.5
Negeri Sembilan	Pantai Teluk Kemang, Batu 8	2314.7
	Taman Tuah Batu	1621.8
	The Regency Tanjung Tuan Beach Resort, Batu 5	459.1
	Kampung Gelam	264
	PD Waterfront	131.9
	Port Dickson District Office	734.4
Melaka	Kampung Portugis	219.4
Pahang	Pantai Cherating	1004.7
	Taman Gelora	497.6
Terengganu	Kampung Teluk Budu	1763
	Taman Geliga	1921
	Pantai Kemasik	308
	Pantai Seberang Takir	935
	Pantai Teluk Lipat	802
	Pantai Paka (Sand Pit)	2557
	Kampung Pak Tuyu	16426
	Kampung Aur	1657
Kelantan	Pantai Kundor-Pantai Cahaya Bulan	952
	Pantai Mek Mas	997
Sarawak	Northeast of Sungai Maludam	2286.5
	South of Tanjung Bungai	3557.1
	Tanjung Paloh	3865.2
	Kampung Semarang	3484.2
	Kampung Santubong	408.2
	Kampung Buntal	1527.7
	Sebangan Bajong (Kampung Sungai Rama)	3465.4
Sabah	Jalan Putatan	841.6
	Kampung Marasimsim	814.8
	Tanjung Tunku	1314.4
Pulau Labuan	Pantai Sungai Pagae near Labuan Crude Oil Terminal	597.2

Coastal Defense in General
Coastal area is a dynamic zone that is highly populated and usually active with economic activity such as harbor, tourism industries and other infrastructure. Apart from it, the coastal area is also home to many marine animals and plants such as mangroves, coral, dugongs and many more. However, developments along the coastal area nowadays have set pressure on the area. Coastal erosion is the common problem that occurs in coastal zone areas. According to Foti et al., (2020), coastal erosion is the consequences of human activities and unbalanced natural changes due to dynamic action such as waves, currents and winds resulting in retreat and sediment loss to the coastal area. In addition, anthropogenic activities such as urbanizations, sand mining, and water resource projects are the main factors to the coastal erosion as these activities disturb and reduce the sediment transport to reach beach area.

The coastal defense structures can be classified into two categories which are hard engineering structures and soft engineering structures. The first category includes structures such as seawalls, groins, jetties as well as breakwaters (Hamakareem, 2012). Meanwhile, the installation of geotextile structures, artificial reefs, hydraulic piling, draining beaches, by-passing and beach nourishment are among common methods that are applied for the soft engineering structures (Atlantic Network for Coastal Risks Management, 2017). Although all these structures play a similar role in protecting the fragile coastal areas, their installation varies according to different needs and situations.

Role of Coastal Defense in Malaysia
Peninsular Malaysia
East Coast Peninsular
The East Coast of Malaysia is the most vulnerable region to erosion compared to the west coast, hence more coastal defenses were built in this zone. In the north part of East Coast Malaysia, Terengganu is one of the states that is most affected during monsoon season. Terengganu has implemented various coastal defenses such as breakwaters, groynes, and rock revetment. According to Ariffin *et al.*, (2019) Kuala Terengganu coastline experiences an annual monsoon season which needs the implementation of coastal defenses to protect the coastal area from erosion. Apart from that, the coastal structures built in this region are also to reduce the impact from coastal development. According to Syakir *et al.*, (2020), multiple coastal defenses were built at about 4km near Kuala Nerus to reduce the impact of erosion due to the development of Sultan Mahmud Airport.

Next, Pahang also implemented the coastal defenses to reduce the erosion problem which mainly due to monsoon season and heavy river discharge from Pahang River. According to Amri Mohd *et al.*, (2018) Pahang coastal region from Cherating to Pekan is vulnerable to the northeast monsoon while the Kuala Pahang experienced level 5 erosion problem due to the high sediment load removal from Pahang river. The implementation of breakwater and rock revetment were actively built especially at Tg Gelang port and Kuala Pahang where these two areas were heavily damaged. Moving down to the south part of the east coast region, Tanjung Piai, located in Johor is prominent with heavy erosion problems due to shipping and coastal development activity. To curb the extension of erosion, various coastal defenses had been used such as geotextile bags, rock revetment, geotextile tube and soft rock revetment. According to Awang, Jusoh, & Hamid, (2014), series of coastal defenses have been implemented since 2003 starting with geotextile bag, seawalls in 2007 to the rock revetment using soft rock in 2010, the erosion problem on to the Ramsar site still ongoing.

East Coast Peninsular
West coast region of Peninsular Malaysia receives lower wave impact from the open sea compared to the east coast region. However, the coastal erosion of the west coast region was reported due to the heavy shipping activity along the strait and mangrove removal for coastal development. According to Shin, Kim, Hakam, & Istijono, (2019), the west coast coastal area is dominated by mangrove habitat. However, since the 1980s, the amount of mangrove along the coast has reduced due to coastal development which promotes

coastal erosion. The implementation of coastal defenses in the west coast is more towards soft engineering to support the growth of mangrove as the natural barrier. In addition, conventional methods such as concrete revetment indeed prevent coastal erosion, however it does not promote natural sediment nourishment. Hence, a soft engineering approach is preferable and suitable for muddy flat sediment in the west coast region. For example, the implementation of geotube breakwaters in Sungai Haji Dorani Selangor reported success because geotube breakwaters are more suitable in areas with smaller hydrodynamic forces.

Next, mangrove replanting efforts are also suitable for the west coast region. Carey island located in Selangor previously experienced extreme loss of mangrove due to anthropogenic activity. This is due to the location of Carey island which is 70km away from Port Klang, also the main factor in the retreat of the mangrove. To prevent the mangrove loss from affecting the erosion, structured mangrove replanting was conducted. According to Bakrin Sofawi, Rozainah, Normaniza, & Roslan, (2017), structured mangrove replanting which used artificial bund and eco wave breaker was found to be successful.

Sabah and Sarawak
The implementation of coastal defenses in Sabah and Sarawak is very limited in the literature. Based on NCES 2015, sandy beaches are common at Sarawak coastline while clay and silt are common soils along the coast of Sabah. Generally, clay and silt are associated with mangrove forests which are the natural protection against waves. However, the mangrove areas are now decreasing due to wave action, natural disaster and human activities which include tourism development at the coastal zones such as resorts and chalets. Among the man-made coastal defense implemented in Sabah is the usage of artificial structure to rebuild the shoreline losses in Selingan island, Sandakan. According to Chen, Saleh, Yap, & Isnain, (2018), Selingan Island is the famous nesting ground for turtles and part of Turtle Island Park (TIP) which experienced beach erosion resulting in the reduction of nesting ground. Hence, Reef balls as artificial structures were invented and implemented to restore the eroded beach. The implementation of the structure has increased sandy beach at the southern part of the island.

Next, similarly to Sabah, Sarawak also less documented the recent coastal structure applied to the state. The recent published of Sarawak coastal defenses was in 2018 which was the impact of erosion at Miri coastal region due to the heavy sediment load from rivers. According to Anandkumar et., (2018), a study was conducted from Baram river to Bungai Beach which covered 11 important tourist spots and commercial beaches at about 74 km to determine the accretion and erosion along the shore. The assessment found out the accretion pattern started after the construction of breakwater, groynes, and rock revetment along the eroded area. 546 acres of eroded area has recovered to 746 acres after the coastal defenses structure implementation.

Applications of Different types of Coastal Defense in Malaysia
The management of coastal issues such as coastal erosion can only be carried out effectively through the usage of suitable methods and techniques. This includes the utilization of coastal protections, comprising both hard and soft defense (Williams et al., 2018). Each of these coastal protections can be used for different applications and purposes depending on the needs and conditions encountered.

Soft Engineering
Nourishment
Beach refills or beach nourishments refer to the addition of sand on the affected or eroded beach in order to increase both beach width and elevation. This soft engineering techniques can be found worldwide primarily in the coastal area with massive development as it functions to reduce the impacts of unmanageable erosion. According to Mangor et al., (2017), nourishment can be grouped into five types which are dune nourishment, backshore nourishment, beach nourishment, shoreface nourishment and profile nourishment (Figure 2). Each types of nourishment have different purpose, for example, dune nourishment is for to strengthen the dune against breaching during acute erosion while backshore nourishment is to strengthen the upper part of the beach (at the foot of the dunes).

Nourishment is one of the approaches that is very flexible and well-suited to adapt to sea level rise as the re-nourishment can be easily adjusted. Through this method, coastal investment as well as beach value can be maintained and retained respectively for the sake of tourism and recreation (Masria et al., 2015). The main advantage of this soft defense is due to its working principle which is highly flexible in allowing sand to shift continuously in response to changing waves and water levels. Besides, the addition of sediment which satisfy the erosional forces can subsequently decrease the impacts of coastal erosion while providing benefits to the adjacent areas through the distribution of sediment by longshore drift. Even so, this technique still cannot be considered as a best solution as it is periodic re-nourishments and not permanent. Other than that, the addition of sediments also can ultimately impose negative impacts to the environment through the direct burial of animals and organisms residing on the beach (Masria et al., 2015). In Malaysia, most of the beaches that has become tourist attraction has done beach nourishment, for example in Teluk Chempedak, Pahang.

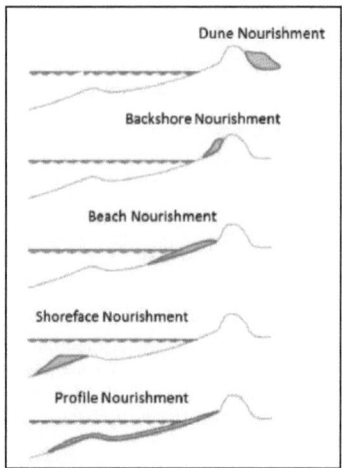

Fig. 2: Different types of nourishment approach

Beach Drain

Beach drain or known as beach de-watering is a system that based on drain in the beach. Based on Mangor *et al* (2017), beach drain helps to increase the beach level near to the installation pipe which directly improving the width of the beach. Beach drain approach is always supported with pressure equalization modules (PEM) systems. It is vertical pipes that arranged forming a matrix along the beach and help in accretion of sand to reduce against erosion. PEM system improve and enhanced beach's ability to drain which make more water can be drain at the top layer of the beach. Thus, more sand deposited rather than washed away by waves. Through this, the groundwater level can be kept at low level (Masria *et al*, 2017). The application of the beach draining system is best for sandy beaches that exposed to tide and sometime moderately exposed to waves. It also good for beach that having only minor erosion to reduce cost needed. However, tit is not suitable to apply beach draining system when the beach is severely damaged due to erosion and erosion that caused by sea level rise. In Kuantan, PEM system were used in beach nourishment back in 2004 to fight against coastal erosion. The evaluation after monitoring process showed that PEM systems and beach nourishment method in Kuantan are successful against, not only minor erosion but the beach's width and level increase as well.

Marsh and Mangrove Restoration

Restoration is a process which aims to return a system to pre-existing condition (Schmitt & Duke, 2015). The definition of marsh and mangrove restoration is referring to the protection of stability of marsh and mangrove platform against erosion and flooding. The mangrove forest acts as a natural barrier to absorb and dissipate wave energy from the sea water. The stability of these platforms will be threatened if the vegetation at the belt were damaged (Mangor *et al*, 2017). Low coastal platform protection required effective management and good public participation particularly the coastal community. The mangrove help as natural barrier to overcome any disturbance or natural disaster i.e. tsunami or storm surge that can affect the coastal properties around the coastal areas. Mangrove restoration can be restored through imposing restriction of activities in the mangrove area, planting new mangrove vegetation re-establish the natural flow in the mangrove area. While for marsh platforms, it can be restored by promoting the natural growth of marsh through construction of siltation traps on the shallow tidal to enhance the marsh growth. In Malaysia, the government has allocated a certain amount of funds for rehabilitation of mangrove under the 9th Malaysian Plan and a small budget was given for conduction R&D related matter (Rahman & Asmawi, 2016). As for the restoration program to be effective, good planning and great site assessment are essential to ensure the survivability of the mangrove belt in the low coastal area. The successful mangrove restoration in Malaysia can be seen at Carey Island where the restoration was supported with used artificial bund and eco wave breaker.

Fig. 3: Natural regeneration of *Rhizophora apiculata*

Hard Engineering
Breakwaters

Breakwater refers to a structure constructed to form an artificial harbour with a basin that is protected from the wave effects. Breakwater can be split into two major types' namely detached breakwater and submerged breakwater. The differences on the application of these structures are the former helps to promote an even distribution of littoral material along the coastline while the latter helps to protect harbours and navigation channels from wave action. Thus, a calm area can be created for ships and tourism activities. By absorbing waves, breakwater helps in reducing wave energy in the leeward part of the breakwater, thus naturally creating salient or tombolo behind the structure which are able to influence longshore sediment transport (Shin et al., 2019). Not only that, current design of breakwater, particularly the submerged type tends to serve another purpose as a multipurpose artificial reef which can indirectly help to develop fish habitat while protecting the coast.

Nonetheless, major challenges in utilizing breakwater as coastal protections are relatively very difficult to build and requires a special design in order to receive an effective result. In the construction of breakwater, there are some parameters that should be considered such as environmental impacts, geotechnical investigation, equipment used to obtain necessary sediment and hydrographic survey. Also, the structures also are quite vulnerable to strong wave action thus require additional structures to support those (Izzat et al., 2018). The common failure in breakwater usually comes from its structural elements and overturning of the wall. In Terengganu, a series of breakwater were built to reduce the erosion impact caused by the construction of airport landing extension which significantly change the sediment transport and greatly erode Pantai Tok Jembal.

Fig. 4: A single attached breakwater in Terengganu

Groynes
Groynes on the other hand are structures that are built perpendicular to the shoreline and work to block parts of littoral drift by trapping and maintaining sand in the upstream areas. Through the usage of groynes, the erosion effects can be diminished as it approaches coastline by altering the current and wave patterns. Groynes may consist of different forms; either emerged, sloping or submerge, and it can be in forms of single or in clusters, known as groin fields. For materials used, groynes can consist of wooden, sheet-pile, concrete, rubble-mound as well as sand-filled (Masria et al., 2015). Different types of materials can be used in different conditions depending on the level of protection required. Besides, this structure is well-favoured to be used particularly in tourism areas as it can build up a beach, resulting in a wider beach that is possible to attract tourists. Even so, the drawbacks of this structure are it requires frequent maintenance with only limited to areas with medium waves. Otherwise, strong waves will penetrate to the cliff face, causing the cliff to be further erode (Williams et al., 2018).

Seawalls
Seawall is a hard structure that was constructed along the shoreline, at the foot of possible dunes. Seawall was built to prevent the shoreline from erosion problems and coastline retreat by protecting the shoreline from wave action and storm surges. Not only that, seawalls also provide other benefits such as opportunities for sightseeing and recreational activities. It is designed to protect the coastline by resisting the force from

the storm surges. A typical seawall usually has sloping structure which either can be smooth slope, stepped-faced or curved faced. Generally, there are three designs of seawall which are rubble-mound structure, block seawall and steel or wooden structure. Sometimes, revetment also was used as a supplement for seawall to slow the scouring process at the toe of the seawall or sometimes it used a single structure at less exposed areas. If the toe of the seawall is severed damaged, it will cause overturning of the wall. This is the main reason why most seawall built have failed. Therefore, it is important to provide protection to the toe during the seawall design process. The construction of seawalls may be costly but with very well planned and design structures, it may be the best solution for coastal protection (Strain et al., 2018; Strain et al., 2020).

Fig. 5: Simple seawall construction at Padang Kota Lama, Penang Esplanade

Revetment
Revetment is passive structure, a shore parallel structure that is constructed which look like seawalls except that revetment is constructed with more horizontal slope, sloppier than a seawall. A seawall is a vertical structure while revetment has distinct slope (Paeniu *et al*, 2015). According to Sadeghi & Al-Othman (2019), revetment is parallel structure to the shoreline to protect coastline from erosions by absorbing and reducing wave energy before they reach to the banks. However, prevention does not protect from flooding and is considered as a supplement to other types of structures such as seawalls or dikes. There are two common groups of revetments which are exposed and buried. As for exposed revetment, there are many types that can be found which are interlocking concrete (<u>Flex Slabs</u>), concrete block, net mesh stone-filled mattresses and geotextile sand tubes.

They added that, there are three important parts in revetment: i) armour layer, important part that protect against wave action, ii) filter zone, block sediment and allow water to pass through and iii) toe coating, protect structure from dislodge and provide necessary support. One of revetment applications can be seen in Sungai Burung, Selangor by using simplified armour unit 'H' or SAUH as concrete revetment for escarpment and bund protection (Department of Irrigation and Drainage Malaysia, 2017). Nonetheless, revetment exhibits a high visual impact on the landscape which can be worse as it can make some beaches inaccessible to people.

Fig. 6: Flex Slab revetment along riverbank in Labuan

Fig. 7: SAUH which are being used in Sungai Burung, Selangor

Fig. 8: Examples of failures of concrete block revetment in Malaysia: Left: scouring of the toe (Penang). Right: By overtopping (Labuan).

Applications of Different types of Coastal Defense in Malaysia
Geotextile tube in sandy coast of Teluk Kalong, Malaysia

Severe erosion of sandy beach has become a major flaw in Teluk Kalong as one of the popular tourism spots in Malaysia. This is due to the effects of rapid wave actions as well as North-East Monsoon in which its wave height can reach up to 1.8 meter and 4.8 meter respectively. Due to these factors, a remedy project under the Public Works Department has been carried out to counteract this issue. This beach restoration project intends to enhance the beachfront value and reduce the erosion level at a minimal cost (Lee et al., 2014). For this project, geotextile tube geosynthetics structures that are frequently used for coastal protection have been utilized. Apart from its low cost and speedy installation, geotextile tube has been applied due to its ability as coastal defense and it only requires simple equipment.

Along the shoreline, a total of 500 m length of the stretches is being covered by the geotextile tubes with a diameter of 3.5 and is located 150 m offshore. Through this coastal protection, it was reported that usage of geotextile tubes is effective in this project as there is an increase of 1.8 m in average for the sediment thickness with estimated accumulation of 87,317 m3 of sediments (Lee et al., 2014). This is because, this beach restoration helps in decreasing the water level leeward of the geotextile tubes and thereby diminishing the incoming wave forces reaching the beach. Thus, incoming dynamic energy that leads to shoreline erosion is reduced, resulting in a low erosion rate (Lee et al., 2014). The differences in the beach condition between before and after the installation of geotextile tubes are depicted in Figure 9.

Fig. 9: Condition of beach (a) before and (b) after the installation of geotextile tubes (2007 – 2008)

Reef balls breakwaters in Selingan Island

The application of artificial structure can be seen in Selingan Island as one of the islands in Turtle Islands Park (TIP) that is continuously being affected by beach erosion. As a tourism spot that offers turtle nesting experience to tourists, erosion due to the impacts of monsoon seasons, extreme events and local coastal processes cause various damage particularly to the habitat and infrastructure. Due to that, Sabah Parks started to collaborate with Reef Ball Foundation for the installation of reef balls as a coastal protection. A total of 290 sets of reef balls were being installed in the southern part of the island by arranging it into three different rows for stability purposes (Chen et al., 2018). The arrangement of the reef balls installed in Selingan Island is depicted in Figure 4. Apart from stabilizing the shoreline through wave attenuation and refraction as a submerged breakwater, reef balls also function as a home to various marine life as well.

Through the application of this coastal structure, the process of sand deposition showed an increment from year 2010 to year 2017 at the southern part of Selingan island. This happens due to the wave that breaks when it comes into contact with the reef balls, thereby reducing the wave energy as the water approaches the shore, decreasing the erosional impact. Other than that, turtle nesting activity also were reported to be active as compared to the condition before the installation of reef balls, indicating that the utilization of reef balls breakwater in Selingan Island can be concluded as effective (Chen et al., 2018). Despite that, the major challenge that involved in this project is that the performance of the reef balls in shoreline protection is highly dependent on the incoming wave energy. Thus, only when the wave energy is low, reef balls are able to function in slowing down the waves and allowing sand to be deposited on these structures or nearby (Chen et al., 2018).

Fig. 10: Arrangement of reef balls in Selingan Island

The state of knowledge on the successes and failures of coastal defenses in Malaysia
Based on what has been reviewed, it is truly understood on what has been done in the implementation of coastal defense in preparing for challenges faced by the coastal areas. Yet, there are always risks of negative impacts if the selection process and development are being ignored by the responsible agencies. Following that, pre- development processes and post- development processes are also significant in ensuring the success of the projects to overcome these challenges. Therefore, one must bear in mind that the selection of the coastal defense structures either hard defense or soft defense must be suitable to protect the shoreline. Generally, a good condition and environment of the coastal areas are essential to access the ability of the coastal defense option to perform as it is required (Chadwick, A., 2020). The causes and effects of the coastal challenges must always be taken into account when dealing with the works involving the littoral movement. This is because the implementation of coastal structures can affect the coastal morphology and result in erosion or accretion of the shorelines. For example, in some cases, the sedimentation pathways may come from offshore sources meanwhile in other cases, these processes may be no longer active. Hence, this review strongly emphasized that the suitability of coastal morphology as the foundation need to be considered in choosing the option and design of the coastal defense.

Furthermore, soft-engineered defense such as sand replenishment would be better implemented as natural defense against coastal erosion and flooding. The approach is considered to be environmentally friendly due to the undisturbed landscape of the beach area in comparison to the hard-engineered defense. However, this approach needs constant maintenance annually by adding sand and shingle as previous deposited beach materials have been washed away by the waves. Nevertheless, when human life and human assets are at risk and need to be protected, the uses of hard elements for a defense may be essential and unavoidable. Importantly, the hard-engineered structures such as groynes, breakwater and sea-gabion are beneficial in absorbing wave energy and protecting the shoreline from coastal challenges. It is worth to note that different options of coastal defense structures have different lifespans and maintenance cost streams. Therefore, comprehensive thoughts must be conducted properly before implementing these coastal protections against the coastal challenges.

CONCLUSION
Coastal erosion can be considered as a natural process in which it is continuously occurring due to the effects from wind, waves, tides as well as currents. However, due to the interference from human activities such as urbanization and heavy development as well as global climate change and sea level rise, coastal

erosion becomes severe and uncontrollable to be resolved. Thus, coastal infrastructures are being used to overcome this issue. In Malaysia, different types of coastal defense have different roles and applications according to specific geographic locations. For the west coast that comprises muddy coast, rock revetment and coastal bund. Meanwhile, coastal defense such as breakwaters, groynes and rock revetment are more familiar to be used in sandy coast of East Coast regions. Additionally, rock revetment, gabion and groynes are mostly used in Sarawak while armour rocks, rock revetment, Labuan block and seawalls are used in Sabah.

Both hard and soft structures are susceptible to different forms of application as well as challenges as a coastal protection. Despite the ability of seawall to effectively protect the coastline by redirecting wave energy back to the ocean water, they are known to be very costly, require large space and highly dependent on seawall's size and shape. For bulkheads that offer protection for the upland area, the challenges involve inability to be used in high energy areas. On the other hand, groynes are applied to reduce the erosion effects through the alteration of current and wave patterns. However, frequent maintenance is needed and preferable to be used only in areas with medium waves. Meanwhile for breakwaters, it is commonly applied for the formation of artificial harbour by reducing wave energy in breakwaters' leeward parts. Nonetheless, the construction process is quite complex and additional structures are usually needed to provide support for breakwaters. As for soft defense, beach nourishment is one of the temporary options for reducing the erosion effects without damaging the beach landscape. The other soft defense which is sand dunes work by trapping and stabilizing blown sand and exhibit low negative impacts, yet it is only applicable in coast with less development.

REFERENCES

Ab Razak, M.S., Suryadi, F.X., Jamaluddin, N., and Mohd Noor, N.A.Z. (2018). Shoreline Planform Stability of Embayed Beaches Along the Malaysian Peninsular Coast. In: Shim, J.-S.; Chun, I., and Lim, H.S. (eds.), Proceedings from the International Coastal Symposium (ICS) 2018 (Busan, Republic of Korea). Journal of Coastal Research, Special Issue No. 85, pp. 631–635. Coconut Creek (Florida), ISSN 0749-0208. Retrieved from file:///C:/Users/user/AppData/Local/Temp/SI85-127.1.pdf

Afshin Jahangirzadeh et.al (2012). Effects of Construction of Coastal Structure on Ecosystem. World Academy of Science, Engineering and Technology. University of Malaya (Kuala Lumpur). Retrieved from http://eprints.um.edu.my/14068/1/v65-136.pdf

Airoldi, L., Abbiati, M., Beck, M. W., Hawkins, S. J., Jonsson, P. R., Martin, D., ... & Åberg, P. (2005). An ecological perspective on the deployment and design of low-crested and other hard coastal defense structures. Coastal engineering, 52(10-11), 1073-1087.

Airoldi, L., & Bulleri, F. (2011). Anthropogenic disturbance can determine the magnitude of opportunistic species responses on marine urban infrastructures. PLoS One, 6(8).

Amri Mohd, F., Nizam Abdul Maulud, K., A. Karim, O., Ara Begum, R., Firoz Khan, M., Shafrina Wan Mohd Jaafar, W., ... Abd Wahab, N. (2018). An Assessment of Coastal Vulnerability of Pahang's Coast Due to Sea Level Rise. *International Journal of Engineering & Technology*, 7(3.14), 176. https://doi.org/10.14419/ijet.v7i3.14.16880

Anandkumar, A., Vijith, H., Nagarajan, R., & Jonathan, M. P. (2018). Evaluation of decadal shoreline changes in the coastal region of Miri, Sarawak, Malaysia. In *Coastal Management: Global Challenges and Innovations*. https://doi.org/10.1016/B978-0-12-810473-6.00008-X

Ariffin, E. H., Sedrati, M., Akhir, M. F., Norzilah, M. N. M., Yaacob, R., & Husain, M. L. (2019). Short-term observations of beach Morphodynamics during seasonal monsoons: two examples from Kuala Terengganu coast (Malaysia). *Journal of Coastal Conservation*, 23(6), 985–994. https://doi.org/10.1007/s11852-019-00703-0

Atlantic Network for Coastal Risks Management (n.d.). Overview of soft coastal protection solutions. Retrieved from https://corimat.net/wpcontent/uploads/2017/03/2_Outil2_56P_EN.pdf

Awang, N. A., Jusoh, W. H. W., & Hamid, M. R. A. (2014). Coastal Erosion at Tanjong Piai, Johor, Malaysia. *Journal of Coastal Research, 71*, 122–130. https://doi.org/10.2112/si71-015.1

Bakrin Sofawi, A., Rozainah, M. Z., Normaniza, O., & Roslan, H. (2017). Mangrove rehabilitation on Carey Island, Malaysia: an evaluation of replanting techniques and sediment properties. *Marine Biology Research, 13*(4), 390–401. https://doi.org/10.1080/17451000.2016.1267365

Buck, P. (2018). *The Design of Coastal Revetments, Seawalls, and Bulkheads*. Pile Bulk Magazine. https://www.pilebuck.com/marine/the-design-of-coastal-revetments-seawalls-and-bulkheads/

Chapman, M. G., & Underwood, A. J. (2011). Evaluation of ecological engineering of "armoured" shorelines to improve their value as habitat. Journal of experimental marine biology and ecology, 400(1-2), 302-313.

Chen, N.-G., Saleh, E., Yap, T. K., & Isnain, I. (2018). Effect of artificial structures on shoreline profile of Selingan Island, Sandakan, Sabah, Malaysia. *Borneo Journal of Marine Science and Aquaculture, 2*(December), 9–15.

Department of Irrigation and Drainage Malaysia (2015). *National Coastal Erosion Study (NCES) 2015. Kawasan-pantai-hakisan-kategori-1.* Retrieved from http://www.data.gov.my/data/ms_MY/dataset/kawasan-pantai-hakisan-kategori-1/resource/ed806db7-d2a2-4173-9989-a015907e8245?inner_span%3DTru

Evans, A. J. (2016). Artificial coastal defense structures as surrogate habitats for natural rocky shores: giving nature a helping hand (Doctoral dissertation, Aberystwyth University).

Firth, L. B., Mieszkowska, N., Thompson, R. C., & Hawkins, S. J. (2013). Climate change and adaptational impacts in coastal systems: the case of sea defenses. Environmental Science: Processes & Impacts, 15(9), 1665-1670.

Firth, L. B., Thompson, R. C., Bohn, K., Abbiati, M., Airoldi, L., Bouma, T. J., Hawkins, S. J. (2014). Between a rock and a hard place: Environmental and engineering considerations when designing coastal defense structures. *Coastal Engineering, 87*, 122–135. https://doi.org/10.1016/j.coastaleng.2013.10.015

Foti, E., Musumeci, R. E., & Stagnitti, M. (2020). Coastal defense techniques and climate change: a review. *Rendiconti Lincei, 31*(1), 123–138. https://doi.org/10.1007/s12210-020-00877-y

Hamakareem, M., I. (2012). Types of Coastal Protection Structures and their Details. Retrieved from https://theconstructor.org/structures/coastal-protection-structures/14020/

Hanak, E., & Moreno, G. (2012). California coastal management with a changing climate. Climatic Change, 111(1), 45-73.

Hawkins, S. J., Burcharth, H. F., Zanuttigh, B., & Lamberti, A. (2010). Environmental design guidelines for low crested coastal structures. Elsevier.

Izzat, I., Im, N., Razak, A., Shahrizal, M., & Safari, M. D. (2018). *A Short Review of Submerged Breakwaters*. https://doi.org/10.1051/mateconf/201820301005

Lee, S. C., Hashim, R., Motamedi, S., & Song, K.-I. (2014). *Utilization of Geotextile Tube for Sandy and Muddy Coastal Management: A Review*. https://doi.org/10.1155/2014/494020

Loke, L. H., Heery, E. C., & Todd, P. A. (2019). Shoreline defenses. In *World Seas: An Environmental Evaluation* (pp. 491-504). Academic Press.

Mangor, K., Dronen, N., Kaergaard, K. and Kristensen, S., 2017. *Shoreline Management Guidelines*. [ebook] Horsholm: DHI. Available at: <https://www.dhigroup.com/upload/campaigns/ShorelineManagementGuidelines_Feb2017.pdf> [Accessed 15 June 2020].

Masria, A., Iskander, M., & Negm, A. (2015). Coastal protection measures, case study (Mediterranean zone, Egypt). *Journal of coastal conservation, 19*(3), 281-294.

MatAmin, Abd., Ahmad, M., Mamat, M., Rivaie, M. & Abdullah, Khiruddin. (2012). Sediment Variation along the East Coast of Peninsular Malaysia. Ecological Questions. 16. 10.2478/v10090-012-0010-6. Retrieved from https://www.researchgate.net/publication/274654555_Sediment_Variation_along_the_East_Coast_of_Peninsular_Malaysia

(Malaysia). Journal of Tropical Biology and Conservation, 14: 83-94. ISSN 1823-3902. Retrieved from https://www.ums.edu.my/ibtpv2/files/06.pdf

Milad Bagheri. et.al (2019). Shoreline change analysis anderosion prediction using historical data of Kuala Terengganu, Malaysia. Environmental Earth Sciences (2019) 78:477, doi.org/10.1007/s12665-019-8459-x. Retrieved from https://www.researchgate.net/publication/334747518_Shoreline_change_analysis_and_erosion_pre diction_using_historical_data_of_Kuala_Terengganu_Malaysia

Ministry of Natural Resources and Environment. (2009). *Coastal Management Activities*. Retrieved from http://www.water.gov.my/activities-mainmenu-184v, 4 November 2014.

Paeniu, L., Iese, V., Jacot Des Combes, H., & De Ramon, N. (2015). 'Yeurt A, Korovulavula I, Koroi A, Sharma P, Hobgood N, Chung K, Devi A. *Coastal Protection: Best Practices from the Pacific. Pacific Centre for Environment and Sustainable Development. (PaCE-SD). The University of the South Pacific, Suva, Fiji.*

Pranzini, E. (2018). Shore protection in Italy: From hard to soft engineering and back. *Ocean and Coastal Management, 156,* 43–57. https://doi.org/10.1016/j.ocecoaman.2017.04.018

Rahman, M. A. A., & Asmawi, M. Z. (2016). Local residents' awareness towards the issue of mangrove degradation in Kuala Selangor, Malaysia. *Procedia-Social and Behavioral Sciences, 222,* 659-667.

Revetment. (2017). Department of Irrigation and Drainage. https://www.water.gov.my/index.php/pages/view/536

Sadeghi, K., & Dania, A. L. (2019). An introduction to onshore structures 'construction.

Sadeghi, K., Abdeh, A., & Al-Dubai, S. (2017). An overview of construction and installation of vertical breakwaters. *International Journal of Innovative Technology and Exploring Engineering, 7*(3), 1-5.

Schmitt, K., & Duke, N. C. (2015). Mangrove management, assessment, and monitoring. *Tropical forestry handbook,* 1-29.

Shin, E. C., Kim, S. H., Hakam, A., & Istijono, B. (2019). Erosion problems of shoreline and counter measurement by various geomaterials. *MATEC Web of Conferences, 265,* 01010. https://doi.org/10.1051/matecconf/201926501010

Strain, E. M., Olabarria, C., Mayer-Pinto, M., Cumbo, V., Morris, R. L., Bugnot, A. B., & Bishop, M. J. (2018). Eco-engineering urban infrastructure for marine and coastal biodiversity: which interventions have the greatest ecological benefit?. *Journal of Applied Ecology, 55*(1), 426-441.

Strain, E. M. A., Cumbo, V. R., Morris, R. L., Steinberg, P. D., & Bishop, M. J. (2020). Interacting effects of habitat structure and seeding with oysters on the intertidal biodiversity of seawalls. *PloS one, 15*(7), e0230807.

Syakir, M., Zulfakar, Z., Akhir, M. F., Helmy, E., Awang, N. O. R. A., Azam, M., Muslim, A. M. (2020). The effect of coastal protections on the shoreline evolution at Kuala Nerus, Terengganu (Malaysia). *Journal of of Sustainability Science and Management, 15*(3), 1–15

Williams, A. T., Rangel-Buitrago, N., Pranzini, E., & Anfuso, G. (2018). The management of coastal erosion. In *Ocean and Coastal Management* (Vol. 156, pp. 4–20). Elsevier Ltd. https://doi.org/10.1016/j.ocecoaman.2017.03.022

Yanalagaran, R., Ramli, N. I., & Ramadhansyah, P. J. (2019, February). Overview of Monsoon Induced Coastal Erosion Disaster in Peninsular Malaysia Based on Mass-Media Reports. In IOP Conference Series: Earth and Environmental Science (Vol. 244, No. 1, p. 012035). IOP Publishing.

NOTES

Printed by Books on Demand GmbH, Norderstedt / Germany